PLANTING CLUES

DAVID J. GIBSON

how
plants
solve
crimes

Planting
CLUES

OXFORD
UNIVERSITY PRESS

OXFORD
UNIVERSITY PRESS

Great Clarendon Street, Oxford, OX2 6DP,
United Kingdom

Oxford University Press is a department of the University of Oxford.
It furthers the University's objective of excellence in research, scholarship,
and education by publishing worldwide. Oxford is a registered trade mark of
Oxford University Press in the UK and in certain other countries

Published in the United States of America by Oxford University Press
198 Madison Avenue, New York, NY 10016, United States of America

British Library Cataloguing in Publication Data

Data available

Library of Congress Control Number: 2022930590

ISBN 978-0-19-886860-6
DOI: 10.1093/oso/9780198868606.001.0001

Printed and bound in the UK by
Clays Ltd, Elcograf S.p.A.

For Stella Lindsay Gibson (1934–2020)

Mum, dog walker, and wildflower lover

PREFACE: AN ENTANGLED BANK

On the morning of Friday, 20 March 2015, Costa Rican National Police officers arrested 42-year-old Sergio Ramirez (not his real name) in Vara Blanca, Heredia for attempting to smuggle 66 orchids out of the country. Rodrigo Arya, Assistant Director of the police in Heredia, said that the plants appeared to be from protected land. But there was a problem; the orchids were not in flower. One non-flowering orchid species looks much like another to a non-specialist, consisting of a few strap-like leaves and some roots. To prosecute a smuggling case, officials need to know the names of each orchid and determine which of the specimens are of rare, protected species, and which might be more common and less valuable. In this case, Ramirez had mixed common yellow *Oncidium* orchids with rare miniature *Lephanthes*. As you can see from Figure 1, it's difficult to tell the different species apart and it requires expert botanical knowledge.

Throughout history, we have been enchanted by the incredible diversity of plants and fungi. Since the time of Plato and Aristotle, we have written records of our quest to understand this diversity. Naturalists, and later plant taxonomists, as they have come to be called, have sought a natural, evolution-based classification of plants. Darwin referred to this diversity as 'an entangled bank' in his 1859 *Origin of Species*.[1] This sense of wonder and appreciation for plants is keenly recorded by many authors, from E. J. H. Corner's classic 1964 book *Life of Plants*[2] to more recent treatments such

Fig. 1. A large collection of Costa Rican orchids seized from an orchid smuggler. When it comes to botanical forensics, it is vitally important to correctly identify the plants.

as Richard Mabey's *The Cabaret of Plants*,[3] Jonathan Silvertown's *Demons in Eden*,[4] and Fiona Stafford's *Long, Long Life of Trees*.[5] I recall reading Corner's book the summer before I went up to Reading University for my botany degree. I was taken aback when during our very first lecture the late Professor Eric Watson[6] asked the incoming 'freshers' which of us had read Corner's book. To my embarrassed chagrin, I was the only one to raise a hand. Nevertheless, in spite of the enthusiasm for flowers and gardening held by the general public and indifferent students, and our increasingly plant-based diets, we persistently ignore their presence in many contexts.

This book addresses the lack of an appreciation for plants and the common inability to identify them (popularly referred to as 'plant blindness'[7]) in the context of the law. Forensic botany is concerned with the use of plants as evidence in legal settings, in both criminal and civil cases. While I will introduce many examples of forensic botany in this book, it is clear from my discussions with law-enforcement officials, and in the writings of some of the most prominent current forensic botanists, that an appreciation of the value of plants in forensics is often lacking. Law-enforcement officials, from crime-scene investigators to forensic laboratory scientists to detectives, rarely have much botanical training. A specialist usually has to be brought in when botanical materials are observed at a crime scene, by which time the scene may be disturbed and key evidence lost. Border-control customs agents are frequently overworked and too swamped to carry out much more than cursory inspections, or have to focus on specific cases and leads. As a result, the number of plant-poaching and -trafficking cases that are prosecuted is much smaller than the extent of these crimes.

The most critical part of any botanical forensic work is the correct identification of specimens or exhibits with evidentiary value, often from microscopic or degraded fragments. And that requires an appreciation of the basic tenets of plant taxonomy, emphasizing the diversity and evolutionary relationships among the major plant and fungal groups. This book is not the place for an exhaustive treatise on plant taxonomy. However, I've woven into the narrative some technical details of plant systematics and botany as needed.

We can trace the origins of plant taxonomy from Carl Linnaeus' 1753 introduction of the binomial naming system, through Darwin's 1859 evolutionary theory of natural selection (and his drawing of the first evolutionary tree in his 1837 *Notebook B*), to modern-day DNA-based classification schemes and the Open

Tree of Life project. The identification of unknown plants from classic morphological or anatomical features remains the bedrock of botanical forensics (Chapters 1–4), but modern methods of DNA analysis (Chapter 5) and chemistry (Chapter 6) are critically important too. Brief explanations of frequently used forensic chemistry and molecular methods are provided in the Glossary.

For forensic botany investigations, it is necessary to correctly identify a plant using its binomial Latin or scientific name to avoid the confusion that comes about from using common or vernacular names. As Juliet said to Romeo, ''Tis but thy name that is my enemy; ... What's in a name? / That which we call a rose by any other name would smell as sweet.'[8] Under this scheme, a rose is in the genus *Rosa*, and there are different species of rose such as wild dog rose, *Rosa canina* and multiflora rose, *Rosa multiflora*. Taxonomists base scientific plant names upon an understanding of the evolutionary relationships among different 'taxonomic units'—that is, species. It is necessary to have a working definition of a 'species' to do this work. The so-called biological species concept is most appropriate here, as we are concerned with identifying and unambiguously naming plants. Under this concept, a species is a group of individuals that can successfully interbreed and produce fertile offspring. Species are maintained by isolating mechanisms between other species including the timing of reproductive events, geographic isolation, and incompatibility genes. The molecular and morphological characteristics that define these groups can delineate and describe different species. We group species into larger taxonomic units called genera (e.g. *Rosa*), which themselves are grouped into families (e.g. Rosaceae), orders (Rosales), and so on up the taxonomic hierarchy. Nowadays, molecular data are used to propose species phylogenies that describe the evolutionary relationship among taxa.

To correctly identify an unknown plant, the botanist needs a firm grasp of a plethora of botanical terms and the facility to use a dichotomous key. My PhD advisor, the late Professor Peter Greig-Smith,[9] always insisted on the correct use of a hand lens (a loupe) to get a close-up view of floral details, and woe betide any undergraduate caught using a plant guide with photographs or drawings instead of the keys and descriptions in a botanical flora (Clapham, Tutin, and Warburg's 1962 *Flora of the British Isles*, 2nd edition was ours[10]). These days, there is increasing use of phone apps, of variable accuracy, to identify plants. Phone apps require little to no botanical knowledge to use.

In most cases, in this book I use 'plants' as a catch-all to include plants (in the strict sense, including algae) and fungi together. But as explained in Chapter 4, for historical reasons, botanists often study both plants and fungi, although they are different organisms.

This book interleaves famous legal cases with aspects of botany that help to solve crimes. I present these ideas in the context of botanical cases, focusing on the admissibility and scientific validity of plant-based evidence presented to the legal profession.

ACKNOWLEDGEMENTS

I have to start by thanking my friend and colleague, the late Stephen Ebbs, who as Chair of the then Department of Plant Biology 'suggested' that I need to teach an additional course as part of my faculty workload. Fortunately, he allowed me to choose what this new course might be. With this encouragement I developed an undergraduate course in Forensic Botany that then prompted the writing of this book.

Deciding to write the book was the easy first step. While researching the scientific content of the material is something I'm comfortable with, turning this into an engaging text that non-botanists and non-scientists will enjoy reading was a different matter. My three previous books have been academic texts. To this end, I'm grateful to Latha Menon, commissioning editor at Oxford University Press, for taking on a scientist to write a trade book, and to Jenny Nugée-Jacob for shepherding the manuscript through to publication.

Many colleagues, friends, and relatives read various chapters, offering valuable suggestions that I sometimes heeded. My daughter, Lacey Gibson, acting in her capacity as a professional editor, and my father, Ken Gibson, were brave enough to read the first draft of each chapter. Other valuable advice and critiques came from Justin Brower, Barbara Crandall-Stotler, Gretchen Dabbs, Alice De Sturler, Buck Hales, Karen Hales, Liam Heneghan, Alicia Jones, Kurt Neubig, Zhe Ren, Stan Smith, Trish Smith, Mark

Spencer, and Ken Thompson. Alicia Jones and Chris Behan set me straight on many points of law. The many students who have taken my Forensic Botany course helped me to better explain the cases I present here. I enjoyed my conversations about forensics with Nathanial Fortmeyer, who enthusiastically provided many new sources to me until the COVID-19 pandemic threw me and others out of the Morris Library. I trawled through many forensic cases that included botanical materials, choosing the most representative of different aspects to present. Names of individuals in the most recent cases, post 1995, are anonymized to protect their identity and avoid additional trauma for relatives of victims. In these cases, the anonymized name is followed upon first mention by '(not his/her real name)'.

I'm grateful for a sabbatical leave during the Spring 2020 semester that allowed me the time to concentrate on writing. I also need to acknowledge the online Twitter community for providing all sorts of leads and ideas. My colleagues at work, especially members of my research lab past and present, provided much-needed botanical inspiration. Finally, thanks to my wife Lisa for letting me get under her feet at home during the pandemic lockdown and 'shelterinplace'.

Carbondale, IL, March 2022

CONTENTS

1

A TREE NEVER LIES

Nay, having since examin'd Cocus, black and green Ebony, Lignum Vitæ, &c. I found, that all these Woods have their pores, abundantly smaller than those of soft light Wood.

Robert Hooke (1665) *Micrographia*.[1] The first known written comparison of the wood anatomy of different trees.

You may recall the high-profile O. J. Simpson murder case in 1995, during which the media reported the forensic evidence in great detail. It glued me to the television. Sixty years earlier, another 'Trial of the Century' involved the kidnapping of the young child of the celebrity Lindbergh family. This case still resonates in the annals of forensic science because it represents a landmark in the use of scientific expert witnesses and was one of the first uses of botanical evidence in a court of law.[2, 3]

A warm spring evening on 1 March 1932, and the Lindbergh family were retiring to bed after enjoying dinner in the drawing-room of their plush Hopewell, New Jersey home. All was good in their well-to-do world until around 10:30 pm, when the family heard frantic cries from Betty Gow, the nurse to their firstborn baby son. Twenty-month-old Charles Jr. had disappeared from the upstairs nursery. The Lindberghs, Betty Gow, and their three servants searched the house frantically. The child's father, Charles Lindbergh, pulled his rifle out of his bedroom closet, called Colonel Henry Breckinridge, a Manhattan lawyer, and then the state police. Rifle in hand, Lindbergh rushed outside and down the drive searching the road. Soon the police arrived on motorcycles and in patrol cars. Within 30 minutes of baby Charles being found missing, radio stations around the nation were broadcasting news flashes, and newspapers were leading their front page with the headline: 'Little Lindy Kidnapped'. The media circus had begun. Anne Lindbergh, baby Charles' mother, recorded in her diary (entry of 5 March 1932[4]): 'It is impossible to describe the confusion—a police station downstairs by day—detectives, police, secret service men swarming in and out—mattresses all over the dining room and other rooms at night.' Reporters quickly swarmed the grounds of the house and had to be banned—only to encamp in the nearby town of Hopewell.

Not only was this a high-profile kidnapping, but the case turned the world of crime scene investigation upside down as it became clear that botanical evidence of wood anatomy would play a key part in solving the crime. For one of the first times, botanical evidence was allowed in the trial of a suspect.

What made this case so sensational to the media was the celebrity status of the family. The father, Charles 'Lucky Lindy' Lindbergh, was an aviator, mail carrier, military officer, author, and environmentalist. He had shot to fame in 1925 after becoming the first person to fly solo, non-stop, across the Atlantic Ocean in his monoplane the *Spirit of St. Louis*. The 33½ hour, 3,600 mile flight from New York to Paris won him the Orteig Prize and the US

Medal of Honor, raising him from relative obscurity to celebrity status.

The other half of the celebrity Lindbergh family was baby Charles' mother, Anne Spencer Lindbergh, née Morrow, an author and well-known aviator in her own right.[5] Daughter of multimillionaire banker Dwight Morrow, US Ambassador to Mexico and US Senator from New Jersey, Anne was a shy, 21-year-old senior at the private, liberal arts Smith College, Massachusetts (where her poet mother was President) when she first met the dashing Charles Lindbergh in December 1927, in Mexico City. After a short courtship, they married on 27 May 1929 at Anne's parents' home back in Englewood, New Jersey. Their first son, Charles Jr., was born on 22 June 1930, Anne's 24th birthday.

The Lindberghs lived in a tall, two-storey house, yet someone had kidnapped the baby from an upstairs nursery without anyone in the house noticing. How could this have happened? The obvious first clue was a wooden ladder found in three pieces abandoned in a clump of bushes 60 feet from the house. By resting the ladder against the wall, the kidnapper had left marks from the ladder top on the whitewashed wall directly beneath the nursery window. Indentations in the ground under the window matched the rails of the ladder. This piece of evidence immediately led police to suspect that the kidnapper had used the ladder to climb into the nursery window. In fact, if you search online for photos of the ladder, you'll see that the police reassembled it and placed it up against the wall below the nursery window—it was exactly the right length. They also found other evidence at the scene. Footprints could be seen leading from the soft earth beneath the window to the bushes where the ladder was found, a ¾ inch

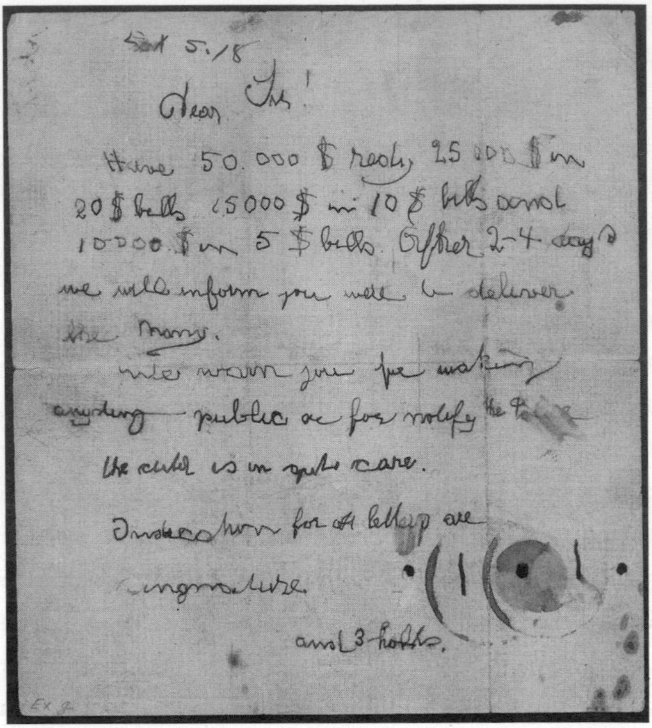
Dear Sir!
Have 50.000 $ redy 25 000 $ in
20 $ bills 15000 $ in 10 $ bills and
10000 $ in 5 $ bills. After 2-4 days
we will inform you were to deliver
the mony.
We warn you for making
anyding public or for notify the Police
the child is in gut care.
Indication for all letters are
singnature
and 3 holes.

Fig. 2. Ransom note left in the Lindbergh baby kidnapping case. Note the unusual signature of two holes, interlocking circles, and an oval.

carpenter's chisel lay in the soil beneath the window, and Charles Lindbergh found a crudely written ransom note in a plain white envelope jammed onto the radiator grill inside the window. The note read (see Figure 2):

> Have 50,000 $ redy. 25 000 $ in 20 $ bills 15 000 $ in 10 $ bills and 10 000 in 5 $ bills. After 2–4 days we will inform you were to deliver the mony. We warn you for making anyding public or for notify the police the child is in gut care. Indication for all letters are singnature and three holes.

The 'singnature' (sic) was two blue interlocking circles with a central red oval.

This note was the first of thirteen ransom notes over six weeks exchanged between the Lindbergh family, their intermediary, a Dr Condon, and a man calling himself 'John' representing the kidnappers. Much like liaisons between cold-war spies and their contacts, there were various drop-off locations, cemetery meeting places, and the passing of notes via unsuspecting strangers (e.g. a taxi driver in one case). An intermediary passed on $50,000 in gold certificates to the kidnapper in return for directions to a location near Martha's Vineyard where the baby could be found. But, despite thorough searches, the police did not find baby Charles, and the kidnapper did not deliver the child. Instead, a lorry driver taking a comfort break in the woods found the baby's body on 12 May 1932, 42 days after the kidnapping. The kidnapper had buried baby Charles in a shallow grave in the woods less than five miles from the Lindbergh home. Injuries to his skull suggested that he died after being dropped by the kidnapper stumbling down the ladder after a rung broke.

An important piece of forensic evidence that helped identify the badly decomposed body was baby Charles' clothing. The dead child found in the woods near the Lindberghs' house was still wearing the nightshirt that his nurse had sewn for him from an old flannel petticoat the night he was kidnapped to keep him warm since he had a fever. Over the nightshirt he was wearing a store-bought sleeveless nightie. But the new, grey, one-piece sleeping garment the baby had been wearing over the other two garments was missing, presumably taken off by the kidnappers, perhaps to use in the ransom negotiations.

Aside from being a home-made ladder used by the kidnapper, what was the forensic value of the ladder? Col. H. Norman Schwarzkopf, who was leading the investigation on behalf of the New Jersey state police, invited Arthur Koehler, a forester working for the US Forest Service in Wisconsin, to take a look. Later in the court proceedings, the prosecution presented Koehler as 'an expert on wood'. The defence counsel, Frederick A. Pope, tried unsuccessfully to disallow Koehler's testimony, claiming that 'there was no such animal known among men as an expert on wood; that it is not a science that has been recognized by the courts'.

In an historical moment for forensic botany, the judge disagreed with the defence, replying 'I deam [sic] this witness to be qualified as an expert', making this one of the earliest cases of botanical evidence being allowed into trial. Koehler had exactly the qualifications and experience to make him an expert witness. With a forestry degree from the University of Michigan and an MS degree from the University of Wisconsin, he was employed by the US Forest Products Laboratory to identify wood specimens, analyse tree growth patterns from wood specimens, test mechanical properties of wood, and advise on milling methods. Indeed, Koehler had testified in an earlier 1923 'Yule Bomber' murder trial of farmer and Swedish immigrant, John Magnuson.[6] In that case, a pipe bomb had been sent two days after Christmas to the home of Wood County, Michigan Commissioner James Chapman, killing his wife Clementine. Magnuson was involved in a land drainage dispute with the local county officials and so became the suspect. After carrying out examination under a microscope, Koehler matched wood shavings of white elm (*Ulmus americana*) in a box that had contained the pipe bomb with wood in the suspect's workshop. The judge sentenced Magnuson to life imprisonment.

Later, after the Lindbergh trial, Koehler said in a 1935 radio interview:

> In all the years of my work, I have become convinced of the absolute reliability of the testimony of trees. They carry in themselves the record of their history. They show with absolute fidelity the progress of the years, storms, droughts, floods, injuries, and any human touch. A tree never lies. You cannot fake or make a tree.[7]

The home-made ladder in the Lindbergh case was constructed in three nested sections like an extension ladder. By examining the wood, especially the grain patterns, Koehler identified three of the six side rails as being constructed from 'North Carolina pine'. He identified ten of the eleven rungs (cross-pieces, or cleats) as ponderosa pine (*Pinus ponderosa*), and one rung and three of the rails from the top of the ladder as Douglas fir (*Pseudotsuga menziesii*). Several birch (*Betula*, probably white birch, *B. alba*) dowels held the three sections of the ladder together.

Identifying a piece of wood as from a pine or fir tree, or another softwood (i.e. a tree in the group of seed-producing woody plants known as gymnosperms that includes needle-leaved pines, firs, spruces, cedars, junipers, redwoods, and yews), is quite easy because of the preponderance of densely packed, water-conducting, fibre-like cells called tracheids (Plate 1).[8] The name is derived from their resemblance to the air-conducting tubes or tracheae in insects. These softwood trees also have large amounts of gum that are secreted into resin canals in their wood. It is this gum that accounts for the familiar stickiness of pine wood. By contrast, hardwoods are in another major group of plants, the angiosperms, which includes trees such as broad-leaved ashes, beeches, birches, elms, maples, and oaks. In addition to the tracheids of the softwoods, the wood of the majority of hardwoods has large numbers of water-conducting cells called vessels, which appear as pores in a cross-section of the wood. As the 'sap rises' in a tree during the spring season, water moves up in the tracheids and vessels. The end walls of individual tracheid and vessel cells abut each other, allowing a continuous tube of water to be literally pulled up the tree as water evaporates from the leaves. New wood elements (tracheids and vessel cells) that are produced by a special ring of dividing cells are generally larger than when these same wood elements arise later in the season. Hence, this difference in size of the new wood elements produced in the spring compared with those produced at the end of the previous year's growing season produces a growth ring. Wood anatomists describe trees that show this pattern of annual or seasonal growth as having 'ring porous' wood. Other trees, such as many tropical species growing in the absence of strong climatic seasonality, show little difference in the size of wood elements through the year and are described as being 'diffuse porous'. An important visual characteristic of wood is the 'grain', a term which refers to the texture revealed on a cut piece of wood because of the pattern of lines from annual growth rings, rays, and knots formed from when branches arise. These grain patterns, along with the colour and density of the wood, allow an experienced forester, carpenter, or woodworker to readily tell what sort of wood that they are looking at. Woodworkers talk about 'fine-' and 'coarse-'grained wood, depending upon the size of the woody elements and the arrangement of rays (arrays of relatively undifferentiated cells that conduct water and solutes) transversely across the stem. Most softwoods, including the southern pines and Douglas fir identified in the Lindbergh case, have a preponderance of tracheids that anatomists describe as fine-grained.

Since Robert Hooke's first microscopic investigation of wood in 1665, it has been known how the details of wood anatomy can reveal the type of wood even more accurately than a visual inspection. My own interest in wood anatomy was spurred by a research project I undertook during my degree, in which I examined samples of fossil wood found in river gravels from the Isle of Wight, off the coast of southern England. These river gravels were deposited during the breakup of the ice sheets during the Quaternary period, about 6,000–7,000 years ago. Through careful sectioning and microscopic examination, I identified samples of European ash (*Fraxinus excelsior*), wild cherry (*Prunus avium*), oak (*Quercus* spp.), and European yew (*Taxus baccata*). The yew samples were particularly interesting and were identifiable because of the absence of vessels and the presence of tracheids (making it a gymnosperm), growth rings, rays one cell in width (uniseriate), an absence of resin ducts (unusual for a softwood), and the presence of unique spiral thickenings in the tracheids (Figure 3). In the British tree flora, the yew is the only softwood to have spiral thickenings in its tracheids, making this identification certain. I also positively identified a yew leaf, providing further confirmation of the presence of this tree in the sediments. You have to be careful, though. Several years later, I was examining thin-section microscope slides that I had made from wood samples cut from tree stumps exposed during an unusually low tide off the Florida Gulf coast. I mistakenly thought that I saw spiral thickenings in the tracheids, and jumped to the erroneous conclusion that I was looking at an ancient yew

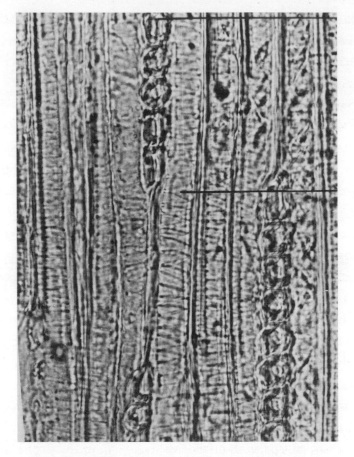

Fig. 3. Longitudinal thin section of fossil yew wood (*Taxus baccata*) showing the presence of diagnostic spiral thickenings in the tracheids and uniseriate rays. Magnification × 175.

forest. There are yew trees along the Florida peninsula, examples of the extremely rare and endemic Florida yew (*Taxus floridana*). But on further examination, the wood turned out to be from the much more common eastern red cedar (*Juniperus virginiana*) that has similar gymnosperm wood anatomy characterized by a lack of vessels, but tracheids with no spiral thickening.

Koehler similarly spent time examining the wood anatomy of the samples from the Lindbergh case. Before the police had apprehended a suspect for the kidnapping, Koehler collected important evidence through painstaking and laborious examination and dogged detective work. He was a true scientist-detective.

Critically, the kidnapper had made one of the ladder rails ('rail 16' from the upper section) from a used piece of poor-grade yellow pine. Koehler would have easily determined that they made this rail from a softwood, and with a fairly rapid microscopic examination that it was a pine. Yellow pine, or North Carolina pine, one of what we today call the southern pines, includes several pine species not readily distinguished by their wood anatomy, including loblolly pine (*Pinus taeda*), longleaf pine (*P. palustris*), shortleaf pine (*P. echinata*), and slash pine (*P. elliottii*).[9] Most likely, Koehler was referring to loblolly pine, as it is sometimes also called North Carolina pine or Arkansas pine. Regardless of species, lumber from southern pines has a wide range of uses, including construction materials, laminates, wood pulp, and paper for newsprint.

Koehler noticed that some of the ladder rails had distinctive machine-plane patterns on their flat and edge surfaces. By examining the rails under oblique light in a darkened room, Koehler could determine from these patterns that the machine plane used to produce the boards the rails were fashioned from must have had eight knives on the top and bottom cutter heads and six knives on the edge cutter. The plane patterns showed that the woodworker must have fed the lumber through the planer at a rate of 230 feet per minute—double the usual rate. Koehler wrote to 1,600 pine mills in the eastern US. Only two factories manufactured planers like this, and they had sold these to 23 mills in the East and South.

After travelling to these mills as part of his investigation, Koehler determined that only the mill owned by M. G. and J. J. Dorn of McCormick, South Carolina, left marks identical to those in the ladder rails. These mill owners had been running their planer speed at 230 feet per minute after they had changed a pulley, and for only a short period of time in September 1929. Between October 1929 and the date of the kidnapping on 1 March 1932, this mill had shipped 45 carloads of 1 x 4-inch lumber matching the ladder side rails to the Great National Millwork and Lumber Company in the Bronx, New York. This shipment turned out to be key evidence in the case.

The gold certificates paid out for the ransom had known serial numbers and started surfacing in New Jersey and later New York. In September 1934, Bruno Richard Hauptmann, a 34-year-old German immigrant carpenter of the Bronx, New York, was found passing off one of the certificates to buy gasoline in New York City. The gas station attendant became suspicious, since the US had gone off the gold standard, and holders of gold certificates were supposed to have exchanged them for 'greenbacks'—paper bills—and ought not to have been using them as currency. The gas station attendant wrote the number plate of the car that Hauptmann was driving in the bill's margin and contacted the police, enabling them to locate and arrest Hauptmann following a car chase close to his home.

The police now needed to link Hauptmann to the scene of the crime, and they had much of the forensic evidence already in hand from examining the ladder. Hauptmann had studied carpentry at trade school before serving as a German army conscript in the First World War. After the war, Hauptmann had trouble keeping a job, and he was arrested, convicted, and jailed for a series of assaults and robberies. Escaping from jail, he made his way to the United States as a stowaway on a cruiser, arriving in Hoboken, New Jersey on 26 November 1923. Although initially penniless, with no legal paperwork or documentation, Hauptmann obtained work initially as a dishwasher and eventually found steady employment as a carpenter.

After Hauptmann's arrest, the wood evidence became even more compelling, identifying him as the person who made the ladder and therefore was likely the kidnapper. On visiting Hauptmann's home, investigators found that a wooden floor board was missing from his attic. Koehler showed that a board used to make rail 16 of the upper section of the ladder had been pried from the wooden flooring in Hauptmann's attic. Nail holes in the ladder rail matched the number, position, and shape of the missing joists. Growth rings in the rail matched up with the adjoining attic boards (see Figure 4). The kerf or width of the saw cuts was 35–37 thousandths of an inch, matching two of Hauptmann's saws in his toolbox. And both rail 16 and the floorboard were of yellow pine. Koehler found sawdust on the ceiling plaster where the board was taken from above to fashion the rail, which matched both the board and rail.

The matching growth rings between the rail and a board in the attic are particularly interesting from a botanical perspective. Koehler was able to testify not only that important parts of the ladder were made from the same type of wood as the floorboards in Hauptmann's attic, but also that pieces of the ladder had been cut from one of these floorboards. Most telling, when the shipments of wood matching the ladder rails had been delivered, Hauptmann had worked as a carpenter with the Great National Millwork and Lumber Company. He also owned a hand-plane that produced

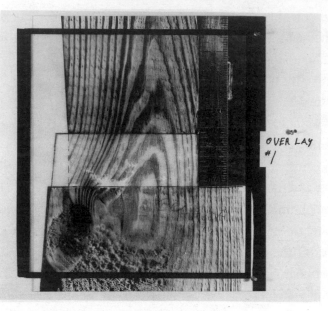

Fig. 4. Comparison of wood grain in ladder rail 16 (top) and attic floorboard (bottom), with overlay showing an artist's rendering of missing piece. Notice how the grain pattern of annual rings extends from the floorboard to the ladder rail. In his testimony, wood expert Arthur Koelher said, 'There are the same number of annual rings in the floorboard as there are in the ladder rail.'

marks on wood similar to those on rail 16 that came from boards in his attic.

Later, Koehler demonstrated these anatomical similarities during the trial by planing some wood in front of the jury in the courtroom. This evidence showed that Hauptmann had access to the boards used to construct the side rails of the ladder, that he could have taken them from the lumber company to make the ladder for the kidnapping, and that he used a board from his attic to make one of these rails using his own tools.

Presumably, Hauptmann's supply of lumber from the Great National Millwork and Lumber Company ran short when he was making the ladder, necessitating the taking of a piece of suitable lumber from his attic.

The case of *State of New Jersey vs. Bruno Richard Hauptmann* began on 2 January 1935 in the Hunterdon County Courthouse, Flemington, NJ. The trial lasted 32 days. David T. Wilentz was lead prosecutor for the State of New Jersey, and Edward J. Reilly was the lead defence counsel. The prosecution called many witnesses, including handwriting experts who testified that Hauptmann wrote the ransom notes. As an expert witness, Koehler introduced the compelling circumstantial evidence that Hauptmann had made the ladder found on the grounds of the Lindberghs' house from lumber he purchased or removed from his attic. After eleven hours of deliberation, the jury found Hauptmann guilty of kidnapping and murdering baby Charles, and the judge sentenced him to death. The state of New Jersey subsequently executed Hauptmann in the electric chair in the state penitentiary on 3 April 1936. He proclaimed his innocence to the end.

The trial had been emotionally difficult for Anne Morrow Lindbergh. After listening to Koehler's testimony, she confided in her diary (entry of 9 February 1935[10]): 'It was a far worse day emotionally than when I testified. That long, long morning of wood testimony: tiny minute points, technical haggling, Vernier scales, how marked, etc. How incredible that my baby had any connection with this!'

The seriousness of the crime that resulted in the kidnapping and death of baby Charles forced the government to act quickly. On 22 June 1932, President Hoover signed into law the United States Federal Kidnapping Act, also known as the Lindbergh Law, making kidnapping a federal offence and, from 1948, a capital crime. Also, several individual states, such as California, passed 'Little Lindbergh Laws' to cover acts of kidnapping that did not cross state boundaries.[11, 12]

Was Bruno Hauptmann really the criminal in this case? Did he work alone? Groundbreaking as the case was for its time, would more modern forensic methods have told a different story? As with many high-profile cases, conspiracy theories abound, with some more credible than others. The official word is still that the crime was solved and the case closed with the 3 April 1936 execution of Hauptmann. However, there are doubts in the minds of some[13] as extensively documented in Anthony Scaduto's 1976 book, *Scapegoat: The Lonesome Death of Bruno Richard Hauptmann*,[14] and *Who Killed Lindbergh's Baby?*, a 2013 PBS NOVA television programme.[15] There is circumstantial evidence that Hauptmann didn't act alone in the kidnapping. There are even suggestions implicating Lindbergh himself. And it has been proposed that Koehler manufactured the evidence linking rail 16 to Hauptmann's attic. Regardless of these doubts, the wood evidence presented by Koehler has stood the test of time, including critical re-examinations of the evidence supporting the verdict. John Douglas, an FBI profiler revisiting the case in the PBS NOVA investigation, concluded: 'If I was working this case, and the police found a piece of that ladder matched wood found at that residence, I would tell the police "Why am I here? Why did you bring me into the case? You got your man."'[15]

* * * * *

Koehler's legacy continues. In 1985, 50 years after the Hauptmann trial, wood anatomist Regis Miller, working in the same USDA Forestry Service laboratory as Koehler, was asked to lend his expertise to the investigations of two crimes.[16] The Alabama State Crime lab needed help. Investigators had found a skeleton at the base of a tree in the Appalachian foothills, north-east of Birmingham, Alabama. There was a shirtsleeve that appeared to be a makeshift noose, knotted around the tree trunk 20 feet above the ground. Did the shirtsleeve indicate a hanging? Was it suicide or a lynching? Forensic anthropologists could only estimate the skeleton had been at the base of the tree for between 1 and 15 years. Why was the noose so far above the ground? Did its height on the tree indicate time since hanging? No! Miller assured the Crime lab investigators that trees grow in height from the top, from the production of new cells in apical meristems at the tips of their branches. A noose, shirt, or anything tied or nailed into a tree does not move up with time. A sign nailed into a tree at eye level stays at that height as the tree grows taller, and as it increases in girth. Miller examined the tree and identified it as a common cucumber tree (*Magnolia acuminata*). He noticed that bark had grown around the shirt where it was tied around the tree limb. Microscopic examination of thin sections of the wood above, at, and below the sleeve showed five annual growth rings constricted where the sleeve tightened around the bark. Someone had tied the shirtsleeve around the tree five years previously. Speculating about why a noose should be so high up the tree, Miller thought that perhaps five years ago the tree had been spindly enough to bend down. The victim could have climbed the tree, bending it down, tied his shirt around the tree to make a noose, and inserted his head into the noose. The tree would have then been let go, springing back up and breaking the man's neck.

A second case presented to Miller was the 'Bundy Tree' case. Hikers in north-east Utah had found an aspen tree (*Populus tremuloides*) with 'Ted Bundy 78' carved into the bark. Aspens are common trees throughout the mountainous regions of the western United States. They have the familiar green leaves that rustle in the wind. Authorities were still looking in the area for bodies from Bundy's serial murder spree in autumn 1974. Could this carving help locate the body of one particular victim, 17-year-old Debbie Kent? Debbie had gone missing on 8 November 1974, after leaving a high school play during the intermission at around 10:30 pm to pick up her brother from a nearby roller-skating rink. She was never seen alive again.

It was thought that Bundy abducted her, killed her, and dumped her body in a recreation area east of the town of Bountiful, Utah. But police had arrested Bundy on 15 February 1978 in Florida so it was very unlikely that it was him who carved into the tree. In a jailhouse interview with Ann Rule, speaking in the third person, Bundy said, 'The business about the tree in Utah with the name Ted Bundy carved in it is bizarre.' Authorities knew his whereabouts in that year, and it wasn't Utah.[17] Perhaps it was a sick joke by someone as a tally count of Bundy's murders. Investigators sent Miller a section of tree trunk that included the carving. After examining thin sections of the bark under a microscope, Miller could see that the knife hadn't penetrated through the thin layer of dividing cells (vascular cambium) that produce the wood inside the trunk. The carving was essentially a superficial wound. But the periderm had been damaged. The periderm ('peri' meaning around or about, and 'derm' meaning skin) is the outer layer of dividing cells that produces the corky bark to its outside that sloughs off a tree trunk as it expands. Eight layers of periderm were next to the scar, meaning that someone had made the carving in 1978 following Bundy's arrest in Florida. The carving wasn't by him, but by someone else. The tree did not indicate the location of a body dump, and so the authorities called off the search for bodies in that location.[16]

* * * * *

In this chapter, I have focused on wood anatomy, which forms one type of botanical evidence. In the coming chapters we will come across other types of botanical evidence that can be important in a forensic investigation. But first, let us look at the legal principles of evidence, specifically the evidence triangle that connects a victim, a perpetrator, and the crime scene. In addition, we will consider the principle that perpetrators or witnesses can easily transfer evidence from one location to another. In many unappreciated cases, this important evidence is botanical.

2

EVERYTHING THAT'S TOUCHED

'Il est impossible au malfaiteur d'agir avec l'intensité que suppose l'action criminelle sans laisser des traces de son passage.' ['It is impossible for a criminal to act, with the intensity that is needed for a criminal act, without leaving traces of his presence.']

Edmond Locard, (1923).[1] Translation courtesy of Lacey Ramirez.

On 9 February 1978, the notorious serial killer Ted Bundy abducted and drove away with 12-year-old Kimberly Leach from the parking lot outside her Junior High School in Lake City, Florida. In his well-planned criminal act, he almost certainly didn't realize that he was carrying evidence with him that would later help investigators tie him to the crime scene.[2, 3] Two months later, after a highly publicized search, investigators found Kimberly's violated and partially decomposed body, dumped in a shed in a wooded area 45 miles away from the scene of the abduction. Plant leaves and soil caught up in the undercarriage of the stolen white van that Bundy was driving, along with fibre evidence and eyewitness testimony, helped convict Bundy for the abduction and murder of Kimberly, his last known victim. Specifically, the identity of the leaves and the characteristics of the soil found on the van when Bundy was stopped for a traffic violation linked him to the scene. The investigators determined that the plants found on his van were also growing at the location where they found the body.[3]

Similarly, convicted murderer Jamie Saffran didn't realize that leaves and plant material from two plants, the umbrella tree (*Schefflera actinophylla*) and Chinese privet (*Ligustrum sinense*), growing in his own garden, would get caught up with the remains and dismembered body parts of his victim, Warren Danzig.[4]

In both cases, investigators brought in professional botanists to examine the botanical evidence. These experts identified the plants that incriminated the two murderers. In the Saffran case, University of Florida botanist John Pipoly had to examine 15,000 plant samples in the Fairchild Herbarium to positively identify the Chinese privet. While Chinese privet is a common invasive plant widely planted in residential gardens, Pipoly had never seen it before in Broward County, where Saffran lived. Therefore, the rarity of Chinese privet in the county helped link Saffran to the body parts of his victim that were scattered across south Florida.

Perpetrators like Bundy and Saffran can inadvertently transport plant fragments from crime scenes, and this provides valuable evidence based upon what's known as Locard's exchange principle. In this chapter, I trace the background to this important principle and consider how evidence in general, and botanical evidence in particular, can be successfully accepted into a court of law. We consider the botanical evidence relevant in a court trial, how this evidence is presented, and by whom—these individuals are the so-called expert witnesses. There are also some important evidentiary standards that must be met, based on well-accepted legal precedents. Jurisdictions use one of two standards in reviewing the admissibility of expert testimony in the United States: the *Daubert* standard or the *Frye* standard.[5,6] Within the US, each state has its own controlling interpretation of either the *Daubert* or the *Frye* standard. If the opposing counsel challenges the inclusion of certain evidence, then the presiding judge may require a hearing to determine whether the evidence has satisfied the appropriate jurisdictional standard. Such legal terminology sounds dry, but considering the principles of science and the scientific method, it is actually quite nuanced and fascinating.

The fourfold purpose of any forensic investigation is to connect evidence at a crime scene to a suspect, the victim, or both, and vice versa. Forensic investigators call these connections the four-way linkage theory (Figure 5). The goal of forensics is to identify and make these connections. This chapter will establish these and other principles of forensic science, starting with Edmond Locard's establishment of the first police forensic science laboratory in Lyon, France.

Edmond Locard was a French forensic scientist.[7] Originally trained in medicine and later in law (he was called to the Bar in 1907), Locard started out as a medical examiner with the French

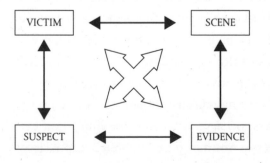

Fig. 5. Four-way linkage theory identifying the connections between botanical evidence, the crime scene, the victim, and a suspect.

Secret Service in the First World War. In that role, he investigated the causes of death of soldiers and prisoners based upon stains and damage to their uniforms. He later served as assistant to criminologist and professor Alexandre Lacassagne, who was a principal founder of the field of criminal anthropology, specializing in toxicology, bloodstain pattern analysis, and bullet markings.

Locard became police prefect in Lyon in 1910 and founded the first police forensic laboratory. Perhaps his superiors were sceptical, as his laboratory consisted of two attic rooms on the fourth floor of the 1835 old Palais de Justice. While the criminal courts in Lyon occupy an impressive, elegant building, visitors could only reach Locard's laboratory through a rear door and up three stories of creaking staircase. Nevertheless, this cramped space became officially recognized in 1912, when it gained worldwide fame as the Police Laboratory of Lyon. Locard become known as the Sherlock Holmes of France. With the help of two assistants, Locard solved many crimes and created his so-called Black Museum, which contained exhibits of murder weapons and other evidence.

However, even before the field existed, Locard was a dogged and obsessive forensic scientist. He experimented on himself to see if fingerprints could be burnt off—finding, painfully, that they could not. Additionally, he conducted a decade-long microscopic study of types of dust ('les poussières' in French) that included plant fragments and seeds, fungi, microbes, insects, decomposition products, and faeces, as well as other dusts such as blood, clothing fibres, and coal, metal, and other mineral dusts. One of his passions was the study of natural history, including cryptogams (seedless plants such as mosses that reproduce by producing spores without flowers), Arctic plants, orchids, and botanical gardens. This passion no doubt influenced his research, as he investigated many types of detritus found at crime scenes, including plant seeds and insects.

Locard wrote prolifically and is regarded as one of the founders of forensic science. His writings included his 1920 treatise, *L'enquête criminelle et les méthodes scientifiques* (*The Criminal Investigation and Scientific Methods*), and it was in this work that he introduced what became known as Locard's exchange principle. He proposed that when two objects come in contact, each takes something from the other or leaves something behind—see the epigraph at the start of this chapter. This principle endures as a key aspect of modern criminalistics.[1]

Locard set the stage for future forensic work with his massive seven volume *Traité de criminalistique* (Treatise on Criminalistics), which includes the 999-page 1931 volume *Les empreintes et les traces dans l'enquête criminelle* (Prints and Traces in Crime Investigation). This treatise included detailed descriptions of different organic and inorganic dusts, methods for collecting and preparing samples for investigation, and 22 example cases ('*Affaires*' in French). Shortly after its publication, a reviewer, Edward Heinrich, wrote that this volume on trace evidence would be 'a new note in the literature of crime detection', having been compiled as a 'masterly array'.[8]

Locard illustrated the value of examining organic dusts in his Observation IX, Affair X (Parisot), in which straw was found on a corpse.[9] On 23 October 1925, police asked Professor Pierre Parisot to help with the case of a dead woman who they had found beaten and strangled under straw in a shed. The killer had tried to make it look like an accident. An autopsy by Dr Morin of the Forensic Medicine Laboratory revealed fragments of straw in the dead woman's eyes and mixed with blood in her stomach. The fragments could only have entered her body while she was still alive, showing that her death was not an accident.

Locard discovered that this straw was from the tropical Kapoc tree (most likely *Ceiba pentandra*). Kapok seed has long, lustrous, yellow-brown fibres, and is known as silk cotton or Java cotton, used as a packing material and as a stuffing for pillows, mattresses, and upholstery. These fibres are an irritant to the eyes, nose, and throat, likely leading to an agonizing struggle for the woman while she was being strangled. Locard notes that this case illustrates the value of examining microscopic particles found on or in bodies.

Locard's exchange principle is the basis of all forensic investigations today. His legacy extends beyond the world of true crime. The Belgian novelist Georges Simenon, creator of the fictional detective Jules Maigret, attended some of Locard's lectures in 1919 or 1920, and Sir Arthur Conan Doyle, author of the Sherlock Holmes detective novels, visited Locard in Lyon. Locard himself credited French translations of Conan Doyle's novels with providing him with inspiration.[10]

Locard was not the first to recognize the value of plants in forensic investigations. While regarded by many as the father of forensic science, Locard himself gave full credit to forensic scientists who came before him. One of these was Hans Gross, a criminal jurist and criminologist regarded by some as the 'father of criminal profiling'. Gross worked as the Examining Justice in the Austrian state of Styria, but he also taught criminal law at Chernivtsi University, Prague University, and the University of Gratz. In 1893, he published *Handbuch für Untersuchungsrichter als System der Kriminalistik* (English title: Criminal Investigations, a Practical Textbook for Magistrates, Police Officers, and Lawyers), which has subsequently been adapted and translated into several languages through at least five editions.[11]

Gross described the value of examining the plant fragments from stomach contents to help identify plants involved in poisoning cases. His work also helped Locard in the identification of wood fragments, including sawdust in many types of crime. For example, Locard attributes his case Observation IV—Affaire X to Gross. This case involved the microscopic identification of the characteristic spines and prickles on the stems of hop plants (*Humulus lupulus*). Gross found large numbers of these spines and prickles on the blade and handle of a knife belonging to a jealous farmer suspected of cutting down his neighbour's hop crop—evidence that helped to convict him.

* * * * *

Does it matter if a professional botanist is not called to testify in a case involving plant materials? Convicted felon D'Angelo Darby may wish that his defence lawyer had called on a botanist to testify at his trial (all names in this case are anonymized).[12] In 2008, Darby was found guilty of the drive-by shooting of Pablo Munio, and botanical evidence linked him to the weapon. He was sentenced to 32 years and 8 months to life. The altercation that led to the shooting started when Darby, who was sitting in the back seat of a green Jeep Commander driven by Rachel Arnold, pulled up at a red traffic light alongside Juan Munio's gold Chevy sedan. Juan's son, Pablo, was sitting in the back seat of the Chevy behind his brother Miguel, who was in the front passenger seat. Apparently, Arnold was laughing at the Chevy, and Pablo stared provocatively at Arnold, who was Darby's girlfriend.

As the lights turned green, both vehicles drove away, but Arnold pulled the Jeep alongside the Chevy. Darby pulled a gun and fired a single shot through the wound-down passenger side window of the Jeep, striking Pablo in the side of the face. Juan Munio then rammed the Jeep, but Arnold sped away. Pursued by police who had been nearby, the Jeep stopped, and Darby and another man exited the back seat of the Jeep and attempted to flee.

Unfortunately for them, two police officers gave further chase and quickly apprehended both men. When Darby was caught by police officer McGrath, he did not have any weapons on him, but he was carrying a black hooded sweatshirt that, critically, had leaves on it. The investigators found a loaded gun dumped in a nearby bush near the area where Darby had jumped out of the Jeep. Forensic tests determined that the gun was used to shoot Pablo Munio. According to officer McGrath, the bush had similar leaves to those on the sweatshirt. During Darby's court case, the prosecution argued that the leaves linked Darby to the weapon.

During cross-examination by the defence, officer McGrath admitted that he was 'no leaf expert'. In other words, McGrath did not know what kind of bush or tree the leaves were from, he did not collect leaves as evidence from the sweatshirt or the bush where the gun was found, and he did not do a detailed comparison of the two sets of leaves. Darby's counsel argued on appeal that he had received an ineffective defence in the original trial, and his council should have called a trained botanist or horticulturist to testify as an expert witness.

The irony is that in this case, it is uncertain how a botanical expert would have helped Darby's defence without the leaves from the sweatshirt or bush to examine. What the botanist would have testified to is mere speculation. The court recognized this argument as speculation and dismissed this point along with other aspects of Darby's appeal. His convictions for premeditated attempted murder, shooting at an occupied vehicle, discharging a firearm from a vehicle, and being a felon in possession of a firearm were upheld and the appeal dismissed.

* * * * *

D'Angelo Darby's counsel suggested in his appeal that his lawyers should have called an expert witness to testify in his trial. According to Rule 702 of the US Federal Rules of Evidence,[13] an expert witness is someone 'who is qualified as an expert by knowledge, skill, experience, training, or education'. This person should be able to help the court understand the evidence, to determine if testimony is the product of reliable principles and methods and based on sufficient facts or data. Additionally, the witness may state opinions within their realm of expertise, they may testify and be cross-examined, and they may be compensated for their time.

The idea of expert witnesses is rooted in English common law. In 1554, Justice Saunders wrote of the case of *Buckley vs Rice Thomas*:

> If matters arise in our law which concern other sciences or faculties, we commonly apply for the aid of that science or faculty which it concerns, which is an honorable and commendable thing in our law. For thereby it appears that we do not despise all other sciences but our own, but we approve of them and encourage them as things worthy of commendation.[14]

While the courts had allowed experts to state facts during trials, this ruling set a precedent by allowing experts to offer their learned opinions. Saunders mentioned cases in which a surgeon and a grammarian had offered their expert testimony. In 1782, in the case of *Folkes vs Chadd*, Lord Mansfield reported that an engineer had offered testimony on behalf of the plaintiffs on the effect of the construction of a bank on the silting up of a harbour.[15, 16]

Nowadays, scientists are frequently called upon to be expert witnesses. If you think you can testify as an expert witness, be prepared to back up your qualifications and your evidence in court. Whether or not you're testifying for the prosecution or the defence, expect the opposing counsel to grill you on the stand to try to discredit you and the expertise that you are offering. Before even getting to the evidence and your testimony, the opposing counsel may ask about your academic degrees, relevant experience with the subject, publications and writings, and whether you have been an expert witness before or not.

Professional crime-scene investigators (CSIs; or scenes-of-crime officers, SOCOs), forensic scientists, and botanists that I have spoken with all say that they are extremely careful in their work, they come to court very well prepared, and, understandably, they get nervous before testifying. If the botanical evidence is strong, then the opposing counsel will try to denigrate the expert witness. In other words, if you can't attack the science, attack the person's authority. As you may recall, in the Lindbergh case, defence lawyer Frederick Pope tried and failed to disqualify forester Arthur Koehler's testimony on the wood used to make the ladder left at the crime scene. Pope did so by suggesting that Koehler was unqualified to make such determinations and that expertise on wood anatomy was not a science recognized by the courts.

However, as British forensic botanist Patricia Wiltshire notes in her 2019 memoir *The Nature of Life and Death: Every Body Leaves a Trace*,[17] even the most skilled lawyer generally knows little about science and even less about botany. As an experienced forensic ecologist specializing in pollen analysis, Wiltshire has honed her courtroom skills over the years through being 'battered and bruised' and engaged in 'mental gymnastics' by opposing lawyers. She points out that her role as an expert witness is to provide 'bullets' that counsel can use in making their case. She also states that most times the evidence that she has presented has led to a confession or acceptance of a plea deal, allowing an expensive trial to be avoided. Indeed it is true that more cases are settled out of court than are prosecuted when details of the forensic evidence, botanical or not, are presented to the defence and the defendant. Nevertheless, it is worth remembering that botanical evidence will be just one part of the forensic evidence and the prosecution's case, making it difficult to know how critical it was. Also, as we learn from Barbara Crandall-Stotler later in this chapter, expert witnesses are not normally privy to other details of a case.

Would it have helped D'Angelo Darby's case if they had called an expert witness to testify on the identity of the leaves on his sweatshirt and on the bush where the gun was found? His defence counsel argued that the arresting officer's lack of expertise compromised his ability to make the match that he claimed. However, a real 'leaf expert' might have testified that the leaves from the sweatshirt and the bushes were not a match. Here, we can never know the truth. I am surprised that the judge allowed the officer's testimony claiming that the leaves matched. In my opinion, the leaf evidence should have been thrown out, given the testifying officer's lack of botanical training.

What happens if the expert witness doesn't do a good job or doesn't have the right qualifications? In 1997, Merle and Nicky Merlaue (not their real names) bought waterfront property along the Columbia River in the western part of the state of Washington. The shoreline along that stretch of the river is steep and rocky, and the Merlaues replaced an old boat dock that was in disrepair with a new one. In 2011, Douglas County issued a notice of violation describing the Merlaues' development of the shoreline as unauthorized under the Shoreline Management Act and the Land Petition Act. Specifically, the county took issue with the construction of a boatlift, a concrete bulkhead, a sidewalk, a patio, a concrete boat ramp, multiple dock floats, a dock ramp, a diving board and slide, grading and retaining walls, and a concrete pad under a hot tub, and with the use of non-native sand. This violation led to fines and a cease and desist order.

Nonetheless, the Merlaues appealed, and as part of their appeal they brought in Tim Roiter (not his real name), a western Washington resident, as a wetland expert. Roiter argued on the day of the appeal hearing that the development carried out by the Merlaues illustrated 'continuity of use' and was best for the environment. But was Roiter a qualified expert witness? He had more than 30 years' experience in wetland evaluation, planning, and permitting at the state and federal level, he was certified as a professional wetland scientist by the Society of Wetland Scientists, and he had additional LEED (Leadership in Energy and Environmental Design) and state of Washington certifications.[18]

Unfortunately, his work in this case appears superficial as he visited the property only on the morning of the hearing, provided no written report, and offered only general observations about the Merlaues' property. More damning, Roiter apparently offered no experience or adequate professional qualifications involving eastern Washington wetlands (five years minimum), as required by the Douglas County Regional Shoreline Master Program in order to offer testimony as an expert witness for this case. The Hearing Examiner noted that 'even if Mr Roiter could be characterized as an expert witness ... [his] purported opinions are not convincing'. The court denied the Merlaues' appeal.

Even as an expert witness, there are limitations on testimony. Again, paraphrasing Rule 702 of the US Federal Rules of Evidence, the expert may offer an opinion if his or her expertise helps in an understanding of the evidence or to determine a fact. The expert's testimony must be based upon the adequacy of pertinent facts or data that are grounded by reliable principles and methods. The expert witness cannot, however, opine on the defendant's mental state or condition related to the crime. Additionally, the counsel must inform the court in advance about an expert witness who it intends to call for testimony; the opposing counsel doesn't like to be surprised. More importantly, it needs time to prepare and possibly bring in its own expert witnesses.

Scientists, and indeed botanists, don't always agree with each other. What can happen if both sides have their own expert witness? In fact, duelling expert witnesses are not uncommon. Expert witnesses are not to be advocates for one side or another, but differing viewpoints about science are common both inside and outside of the courtroom. Scientists are used to disagreeing with each other on interpreting their data. Usually, such academic feuds are carried out within the generally polite confines of academic literature, though they can get difficult and unpleasant, especially when there are big scientific egos involved. Yet in courts, differences in opinions between experts take a different course.

For example, the decomposition of leaves under a baby's body played a key role in the 2011 trial of a woman who was charged with the first-degree murder of her two-year-old daughter.[19, 20, 21] Because of the severity of the crime, prosecutors were seeking the death penalty for the mother. As part of the trial, the defence and prosecution each called a different botanist as an expert witness. The botanists disagreed and offered opposing testimony. The timeline was confusing, as the mother made various conflicting claims, including that the daughter's father had kidnapped or drowned the child in a swimming pool, or that the babysitter had taken the baby.

The child's grandfather was the last person other than the mother to see the child alive when the mother and daughter left his home with backpacks in the early afternoon of 16 June 2008. The police arrested the mother on a charge of child neglect the day after her mother reported the child missing on 15 July. It wasn't until December 2008—six months after she disappeared—that a utility worker found skeletal remains of the child dumped in the woods near the mother's house in Orange County, Florida. Investigators were able to identify the body as it was wrapped in a Winnie-the-Pooh blanket in a tan-coloured canvas laundry bag from the mother's household.

Weeks before the baby's body was found, cadaver dogs picked up the scent of human decomposition in the trunk of the mother's car, but that didn't help them find the body. The prosecution argued that the mother had suffocated her child with duct tape, left the body in her car until the decomposing odour became too strong, and then dumped the body in the woods. Unfortunately, the forensic lab contaminated any potential DNA on the duct tape that the police found around the child's skull. All of this led to a key question—how long was the body in the woods?

The defence brought in forensic botanist Jane Bock from the University of Colorado as their expert witness. Bock is a very experienced and well-qualified forensic botanist with particular expertise in plant morphology and ecology. When the child's remains were found, vegetation and tree roots were growing through her body. Bock testified that the child's body could have been in the woods for as little as two weeks, a much shorter time than that claimed by the prosecution. During that time, Bock argued, roots could have grown through the decomposing body and laundry bag. Bock explained that she established her estimate of how long the body had been in the woods from examining photographs and visiting the crime scene. Specifically, she based her assessment on the 'pattern of the leaf litter'.

However, on cross-examination, Bock agreed that the remains could have been in the woods longer than her two-week estimate. Inadvisably straying away from her area of expertise, Bock suggested that a dog or a coyote could have buried a bone of the child's found in the mud nearby. At this point, the prosecuting attorney pointed out that there are no coyotes in Florida.

Bock contradicted the state's expert witness, David Hall, when asked about vegetation found in the mother's car. Hall, a private consultant, had written a report suggesting that vegetation fragments, mostly leaf fragments, helped place the mother at the crime scene where the child's body was found. Hall had identified the leaf fragments in the mother's car as coming from a camphor tree (*Cinnamomum camphora*). This ornamental tree is native to China and Japan, but was introduced into Florida as far back as 1875 and is now listed as a Category 1 exotic species by the Florida Exotic Pest Plant Council.[22] In response to questions from the defence counsel, Bock pointed out that there were no camphor trees at the crime scene.

Hall had also reported that plant root growth at the crime scene showed that the child's body had been there for four or more months. Defence attorney Cheney Mason filed a motion asking that Hall's testimony in his report be thrown out because his 'opinion is the result of his personal experience as a result of watching roots grow since he was ten years old', and that the application of appropriate scientific method to support his claims was lacking. In response, Chief Judge Belvin Perry Jr ruled that Hall's testimony and opinion would be admissible. In doing so, the judge was indicating that an evidentiary *Frye* challenge was not needed since Hall's testimony was based upon opinion that the jury could judge (we will look into *Frye* and the pure opinion rule in detail at the end of this chapter). Often, when counsel raises potential weaknesses in an expert's opinion testimony, the judge will admit it anyway on the grounds that the weaknesses go to the weight, but not the admissibility, of the testimony. In this case, Bock's opposing testimony, based upon her botanical expertise, along with the defence motion, helped counter arguments presented by the prosecution.

The judge acquitted the mother of murder and manslaughter charges, and she was only found guilty of a misdemeanour of lying to the police. The case remains unsolved, and prosecutors can never try the mother for it again because of the principle of double jeopardy; that is, no one can be prosecuted twice for essentially the same crime.

The personal experience of botanists as expert witnesses is eye-opening to those of us used to the 'ivory tower' of academia. A colleague of mine, Barbara Crandall-Stotler, an international authority on bryophytes (a plant group including mosses and liverworts of small evolutionarily ancient green plants that reproduce via spores or clonally, and lack flowers or vascular systems), and an emeritus professor in the School of Biological Sciences at Southern Illinois University, Carbondale, recounted to me her experience testifying at a local murder trial.[23] The case involved charges of murder against defendant Army Sergeant Charles Ashley, aged 23.[24] Prosecutors charged Ashley with murdering Melinda Buchanan, aged 24, and her two daughters, Jennifer aged 2, and Rachael aged 4, on 1 February 1987. Ashley had been on leave from his post at Fort Hood, Texas to attend the wedding of his sister in Robinson, Illinois. He had been out drinking on the evening of 31 January during which he visited the Buchanan house, allegedly to get the phone number of Melinda's husband Ben, who was in California.

At 5:30 a.m. on the morning of 1 February, a passer-by notified the local volunteer fire department that the Buchanans' house was on fire. Firefighters found the three victims dead in the house after they extinguished the fire. The two children had died of smoke inhalation, but a pathologist determined that the mother had died from strangulation before the criminal had started the fire. The criminal had deliberately set the fire in three places around the house: one in the master bedroom where firefighters found Melinda Buchanan's body, one near the children's bedroom, and another in the bathroom.

Firefighters found Melinda Buchanan with her underwear pulled below her hips. At the trial, Glenn Shubert, a hair and fibre expert, testified that three pubic hairs found on Melinda's underwear were consistent with samples from the defendant. This hair evidence ties the defendant to the murder victim. Other evidence linking the defendant to the crime included four fragments of Buchanan's head hair found on a white T-shirt with a 'USS Saratoga' logo under the driver's front seat in the defendant's car. Witnesses confirmed that the defendant was wearing the T-shirt the night of the crime. Finally, investigators found a black jacket belonging to the defendant in the Buchanans' master bedroom.

However, it was the botanical evidence that helped tie everything together. The investigators found 'moss-like' botanical fragments throughout the crime scene: on Buchanan's underwear and a tank top that she was wearing; under her fingernails; on a pair of blue jeans, a pink sweater, and a sock found on the floor of her bedroom; on the black jacket belonging to the defendant; and on the T-shirt and a blue-and-white bandana found in his car. Personnel from the State of Illinois Bureau of Forensic Sciences collected these exhibits from the scene and catalogued them along with other evidentiary items.

A representative of the Illinois State Forensic Lab delivered the botanical samples to Barbara Crandall-Stotler. In her research laboratory, Crandall-Stotler examined the fragments under a dissecting (stereo) microscope. She then mounted leaves and branch tips of the moss fragments in water onto microscope slides for closer examination with a compound microscope.

Crandall-Stotler took detailed notes and photographs of each sample, using a camera mounted onto the compound microscope, and she then dried and returned each sample to its numbered packet. A dissecting microscope allows examination of whole fragments, in light shone down from above, at up to a $100\times$ magnification. The two eyepieces allow scientists to view a three-dimensional image of the object. Additionally, scientists use a compound microscope in which a light from below passes through the specimen to examine thin samples through multiple lenses up to $1,000\times$ magnification and to examine cellular detail.

There's always excitement and anticipation when putting a botanical specimen under the microscope for the first time. Will it be something familiar or something unusual? Here, Crandall-Stotler readily determined that the botanical material collected from the crime scene was, with one exception, branch tips and leaves of brown, horticultural *Sphagnum*, commonly known as peat moss, intermixed with dark potting soil and horticultural vermiculite.

The leaves of *Sphagnum* plants have a unique arrangement of two types of cells. There are relatively large, dead, water-containing hyaline cells, which have pores in their cell walls, allowing the leaves to soak up 16–26 times as much water by weight as the plant itself. And between the hyaline cells are small, green, living chlorophyllose cells that undergo photosynthesis. The leaves are arranged in characteristic clusters of branches or fascicles, with two or three spreading branches, and two or four hanging branches. A trained botanist could not confuse *Sphagnum* with anything else. When dried, *Sphagnum* is a common component of ornamental potting mixes because of its ability to retain water and nutrients.

As expected with this moss, there was some variability in morphology of the leaves and among cell sizes, but nothing to dispute the conclusion that all were from the same species of *Sphagnum*. The exception was a single stem of the green woodland moss *Amblystegium* found on Melinda Buchanan's blue jeans. Crandall-Stotler submitted her report to the Division of Criminal Investigation in Effingham, Illinois on 20 May 1987, concluding that all the *Sphagnum* fragments came from the same source: likely house plant potting mix. Crandall-Stotler returned the exhibits to the State Forensic Lab, maintaining the chain of custody at all times. She also gave the microscope slides and photographs taken of the samples to the Forensic Lab. It is important to know that at this time, Crandall-Stotler knew little more about the case beyond the plants she examined. This limited knowledge about the broader aspects of a case is important and typical in forensic examinations, to avoid bias.

What role did this botanical evidence play in the trial of murder suspect Charles Ashley? Along with other forensic evidence, including the hair samples and clothing, the botanical evidence helped link the suspect to the crime scene. Crandall-Stotler gave testimony at the trial. As is usual to avoid bias, she was sequestered during trial and only in the courtroom for her own testimony. However, experts are sometimes permitted to sit in trial and observe the testimony of other experts—and then comment on that testimony when they themselves testify. Nevertheless, during the testimony phase in this case, the prosecution and defence questioned Crandall-Stotler about her credentials to serve as an expert witness, the methods used to examine the samples, and her conclusions. These conclusions included showing the jury the Kodachrome slides that she had taken of the samples from the victim's clothing and the T-shirt and bandana from the defendant's car.

It was only later, after the trial, that she learned that there was a knocked-over potted plant on the floor of the bedroom where the victim was found. The *Sphagnum* fragments found under the fingernails and on the hand of the victim and on the clothing of the suspect suggest that the two of them had pulled over the potted plant during the struggle that led to her death. Paradoxically, the CSIs did not collect samples of the potting mix; perhaps they had not noticed it. If the potting mix contained the same *Sphagnum* moss, soil, and vermiculite that Crandall-Stotler identified in the exhibits she examined, then there would have been an even better link between the suspect and the crime scene.

Nevertheless, the jury of ten men and two women found Charles Ashley guilty of three counts of murder and aggravated arson. The judge sentenced Ashley to natural life imprisonment for the murder charges and 50 years for the arson charges. He remains incarcerated in Menard Correctional Centre, Illinois, ineligible for parole.

* * * * *

In the trial of the Florida mother accused of murdering her child described earlier, we saw how the judge ruled that prosecution expert witness David Hall's testimony and opinion of the botanical evidence would be admissible, and not excluded subject to a *Frye* challenge. In addition, the judge also ruled that testimony by prosecution expert Dr Arpad Vass on the 'odour signature' of a carpet sample from the mother's car boot met the *Frye* standard for admissibility. Dr Vass had testified that the odour was consistent with that of human decomposition.[25]

The *Frye* challenge is one of the minimum legal standards that scientific evidence has to meet in many US courts. We saw already that allowing certain experts to offer their opinion into testimony to advise the court has its origins in English common law dating back at least to the 14th century. The jury must make the ultimate findings of fact, but they need help. These expert witnesses, or skilled witnesses, can provide expert testimony, but it is subject to several tests or challenges. In the United States, the most important standards are the *Frye* and *Daubert* challenges, named after two precedent-setting cases.

In 1923, James Alphonso Frye argued to the US Court of Appeals in the District of Columbia that the court should disallow his confession to the 20 November 1920 murder of a Dr Robert Brown.[5, 26] Earlier, a judge had convicted Frye of second-degree murder following the shooting of Brown, despite the absence of any other evidence. At issue in the appeal was the exclusion of expert witness opinion of psychologist William Marston on the use of a systolic blood pressure detection test. Marston claimed that this test provided a measure of the relationship between a subject's blood pressure, emotions, and mental competency (i.e. a then-new type of lie detector or polygraph test). Marston claimed that the test would show that Frye had falsely confessed to the murder of Brown. Apparently, he made this confession to detectives in exchange for them dropping some unrelated burglary charges.

The prosecution argued that the blood pressure test didn't have the required standing or scientific credibility from the psychological and physiological authorities. The court accepted the prosecution's argument and affirmed the denial of this testimony and the appeal. As the confession was the only evidence linking Frye to the murder, his conviction stood.

By rejecting the appeal, the court set an objective standard for admissibility of expert testimony, writing that 'the thing from which the deduction is made must be sufficiently established to have gained general acceptance in the particular field in which it belongs'.[5] Many US states and courts worldwide subsequently adopted this *Frye* or 'general acceptance' standard. It means effectively that the word of a single expert may not be enough; rather, the consensus of experts in a field needs to be established. The *Frye* standard has now been superseded in many US courts with the more detailed 1975 Federal Rules of Evidence (FRE; specifically, Rules 701–706 on 'helpfulness' and 'relevance' standards).[27] However, the 1923 ruling dealt the death knell to the use of polygraph tests that still stands today as inadmissible evidence in most jurisdictions.

The courts established a second precedent-setting evidentiary case in 1993. In this case, lawyers on behalf of plaintiffs Jason Daubert and Eric Schuller sued drug manufacturer Merrill Dow Pharmaceuticals, claiming that the morning sickness drug Bendectin was causing birth defects.[28] Jason Daubert and Eric Schuller had been born with serious birth defects after their mothers had taken Bendectin. This drug is a cocktail of deoxylamine succinate (an antihistamine), pyridoxine hydrochloride (vitamin B6), and dicyclomine hydrochloride (an antispasmodic) as active ingredients.

Daubert and Schuller's claim against Merrill Dow Pharmaceuticals was not the first case alleging problems since the U.S. Food and Drug Administration (FDA) approved the drug in 1956; by 1983, there were more than 300 pending lawsuits. Doctors widely prescribed Bendectin, and more than 33 million women worldwide used the controversial drug. The first widely publicized lawsuit was the 1977 Mekdeci case, in which the plaintiffs argued that Bendectin caused Poland syndrome, a birth defect that caused a deficit in the pectoralis muscle. The outcome of that trial, which involved the testimony of many expert witnesses, was inconclusive. Although the plaintiffs prevailed, the court awarded them only a small amount of damages ($20,000) and no fees. The Mekdeci trial sent a warning message to the drug manufacturer, but did not stop the company from marketing Bendectin.[29]

The later *Daubert vs. Merrill Dow Pharmaceuticals* trial and the subsequent 1993 decision by the US Supreme Court were not only more conclusive than the Mekdeci trial but had far-reaching and worldwide implications for expert witness testimony. The court considered the credibility of the raft of scientists brought in as expert witnesses for both sides. In particular, the court declared inadmissible the expert testimony of the plaintiff's main expert witness, William McBride. He based his testimony on *in vitro* and *in vivo* animal studies, pharmacological studies, and the reanalysis of other published studies using methodologies that had not yet gained acceptance within the general scientific community. Indeed, McBride was later found to have falsified his research, and he was struck off the medical registry in Australia. The appeals court ruled in favour of Dow, who nevertheless withdrew Bendectin for a while before making it available again under new names.[30,31]

The *Daubert vs. Merrill Dow Pharmaceuticals* decision established that the judge acts as gatekeeper to determine admissibility of opinions and testimony of an expert witness. It recognized that the FRE superseded the *Frye* standard and established the following new standards specifically that the judge should determine for the court in a particular case:[32]

- whether the theory or technique concerned can or has been tested;
- whether it has been subjected to peer review and publication;
- its known error rate;
- the existence and maintenance of standards controlling its operation;
- its widespread acceptance within the relevant scientific community (as per *Frye* but not directed towards settling a case).

Expert witnesses can offer opinion within the realm of their expertise, besides helping the court understand scientific facts. Jane Bock and David Hall offered contrasting testimonies in the trial described earlier about how long a child's body had been in the woods. Their testimony was based upon their opinion of the condition of the leaves under the body and not any scientific tests. They had already established their credibility as experts; they had satisfied the court, and so opposing counsel did not subject their opinions in this part of their testimonies to a *Frye* challenge. Rather, the presiding judge, Belvin Perry Jr, considered their testimony was based upon the 'pure opinion rule' evidentiary concept relevant for expert witness testimony. The pure opinion rule appears to be a way around *Frye's* otherwise strict general acceptance standard. The expert could give their pure opinion if an opinion can be viewed as based on their experience and training, but not directly related to a new scientific theory or test (which would then be subject to *Frye*). This concept is derived from the 1993 *Flanagan vs. State of Florida* appeal of a child abuse case, where a psychologist offered her opinion on the competence of the defendant based upon her personal experience and training. In this case, Justice Grime wrote:

> Pure opinion testimony, such as an expert's opinion . . . does not have to meet Frye, because this type of testimony is based on the expert's personal experience and training. While cloaked with the credibility of the expert, this testimony is analysed by the jury as it analyses any other personal opinion or factual testimony by a witness.[33]

The expert witness testimony and evidence admissibility challenges of *Frye* and *Daubert*, the FRE, and the pure opinion rule are rulings from US courts, but differ in their applicability from one state to another. Most criminal cases are tried in state courts, and each state has its own evidence codes and doesn't follow the federal code. Since 1993, *Daubert's* interpretation of Federal Rule of Evidence 702 has been the controlling standard; prior to that time it was *Frye*. The majority of US states follow the *Daubert* standard in their evidence codes, but a minority do not. At the time of the 2011 child murder case discussed earlier, Florida was a *Frye* jurisdiction. In 2019, Florida became a *Daubert* jurisdiction. In contrast, Illinois, where I live, remains a *Frye* jurisdiction.

These standards are also not binding in other countries which, nevertheless, have comparable standards. In Australia for example, the 1984 Supreme Court Appeal of the forgery case *The Queen vs Bonython* established a two-part test based on *Frye* determining (a) whether the court needs expert witness(es) to explain the subject matter to the jury, and (b) whether the subject matter of the opinion forms part of an organized and reliable body of knowledge.[34] Researchers now also cite *Bonython* as an authority in cases and expert witness admissibility guidelines in the United Kingdom. Generally, the UK and the US share similar principles governing expert witness testimony, with a few differences in the details.

A botany case focused on adjudicating potential damages to a crop following unintentional herbicide drift illustrates the nuances of a court in applying *Frye* and *Daubert* standards in admissibility of expert witness testimony. In 1996, Illinois farmers Frank and Lydia Lewis (not their real names) alleged that an employee of the Cropmate Company was negligent in spraying the herbicide 2,4-D (2,4-dichlorophenoxyacetic acid) on farmland next to a 50-acre tract of land rented by the Lewises, damaging their watermelon (*Citrullus lanatus*), cantaloupe (*Cucumis melo*), and pumpkin (*Cucurbita pepo*) crops.[35] 2,4-D is a widely used synthetic herbicide for controlling broadleaf weeds by causing uncontrolled, lethal cell division and growth in sprayed plants. Most grasses metabolize 2,4-D differently to broadleaved plants and so are resistant to its effect. Although scientists in biotechnology companies engineer some crops such as corn (which is a grass anyway) to be resistant to 2,4-D, it is best applied prior to sowing or crop emergence.

The Lewises successfully sued for damages. On appeal, the defendants argued that the court erred because expert testimony offered by the plaintiffs should have been excluded, as it did not meet *Frye* or *Daubert* standards. The testimony in question—a visual assessment of any damage to the Lewises' crops caused by 2,4-D exposure—was provided by three experts: a seed salesman with herbicide experience, an extension specialist and weed science student, and an academic research scientist with experience of 2,4-D effects on cucurbits. These experts used a 'comparative symptomology' technique, which is generally accepted by the scientific community.

Affirming the original court finding, the appeals court ruled that the *Frye* standard did not apply because the visual damage estimates did not constitute 'scientific' evidence, the methods were sound, and they were not novel. Courts in the state of Illinois had not yet adopted the *Daubert* standards, and the appeals court did not think they applied in this case anyway. Rather, Illinois courts relied on a '*Frye* plus reliability' standard, of which the 'community' of experts proffering evidence adequately addresses the second reliability component here. In the end, the appeals court affirmed the trial court's judgement.

Can we profile a botanical expert witness? It is interesting to consider what kind of botanical experts are qualified to provide testimony in cases involving plants. There are various botanical areas that can come into play, depending upon the evidence (plant taxonomy and systematics, anatomy and morphology, genetics, ecology, palynology, diatomology, phycology, mycology, and so on). We will look at the forensic evidence that these areas of botany provide in the following chapters.

Recall that Arthur Koehler had very specific expertise on wood anatomy that made him the perfect expert witness for the baby Lindbergh case (Chapter 1). And Barbara Crandall-Stotler, a bryologist, was the perfect expert for examining the *Sphagnum* peat moss in the Charles Ashley case. Finally, the opposing counsels brought in Jane Bock, an ecologist, and David Hall, a plant taxonomist, to testify on leaf decomposition patterns in the child murder trial. Crandall-Stotler, Bock, and Hall did not have to face *Frye*, FRE, or *Daubert* challenges as their credentials were impeccable (although challenged by opposing counsel), and the methodology they used was straightforward.

Indeed, Bock's and Hall's testimonies about leaf decomposition would more appropriately fall under consideration of the pure opinion rule, since neither made any measurements nor conducted any tests. DNA technology is a different matter (Chapter 5); it is rapidly changing, with the introduction of new methods such as next-generation sequencing, which are likely subject to *Frye* and *Daubert* challenges. Indeed, the presiding judge required a *Frye* hearing when the prosecution introduced the first plant DNA evidence in the Maricopa case.

Botanical training is conducted almost exclusively in universities, but it is declining as botany departments are disappearing, often becoming absorbed into broader schools of biological sciences. (This change has happened at my institution, after it being a separate academic unit since 1929—almost 100 years.) In many countries, forensic science labs don't have botanists on staff, CSIs are untrained in botany, and botanical evidence is contracted out to private investigators or university faculties. For example, the closing of the Forensic Science Service in the UK in 2010 has meant that forensic botany evidence and investigation are in the hands of a few, albeit very competent, individuals. Nevertheless, in the following chapters we will consider how the different types of botanical materials can provide important forensic evidence.

3

GETTING CAUGHT UP

Brambles often grow in places where people have done bad
things to other people.

Mark Spencer, forensic botanist[1]

Edmond Locard, the 'Sherlock Holmes of France', was stumped.
He had a dead body and no clues to help him find the killer.
Police had found the body in the countryside outside Lyon, with
a knife in its heart. The local gendarmes had thought only to
retrieve the body and had trampled the crime scene, obliterating
any footprints that might have helped track down the assailant.
Yet Locard was persistent, and when the local gendarmes had
rounded up some vagrants a few days later, one of them, a railroad worker, had a bloody stain on his jacket that caught Locard's
attention. Upon closer examination of the jacket, Locard found a
seed of what he first thought was a dandelion plant (*Taraxacum* sp.).
Nevertheless, that finding didn't help Locard as dandelions have
a worldwide distribution and are very common in many habitats.
However, after he took a closer look, Locard realized that it was
actually the seed of a different and, in fact, relatively rare plant
in the dandelion family (the Compositae, now usually called the
Asteraceae).

Worldwide, there are more than 32,000 species in the Asteraceae family, with many of them growing in the area of Lyon,

France, so there were still plenty of species for Locard to choose from. Locard doesn't tell us which Asteraceae plant it was, but he recalled that there was a tuft of the same plant species growing close to the corpse. Commonly, the dispersal unit of species in this family comprises a small, hard, one-seeded fruit (the achene) below a ring of hairs or bristles (the pappus). These hairs allow the seeds to be readily dispersed and get caught up in materials. It was a seed of one of these plants that had become attached to the clothing of the killer, the railroad worker. As Locard noted, 'Ce détail, d'apparence infime, c'est venu résoudre le problème: l'assassin est pris.' (This seemingly minute detail solved the problem: the murderer is caught.)[2]

Locard's case, his 'Observation XVII—Affaire X—Traces de végétaux,[3] illustrates several important points, namely the value of a close examination of a suspect's clothing for small clues (dust or 'les poussieres' in Locard's terms), of taking a close examination of the crime scene (noting the nearby plant species), and of the botanical knowledge to know or be able to identify what you are looking at. By using botanical evidence to link a suspect to the crime scene, this case also provides a perfect example of Locard's exchange principle and the four-way linkage theory (Figure 5) of forensic science.

In this chapter, I will describe several cases that illustrate different aspects of Locard's exchange principle using macroscopic plant data—the utility of plants, plant parts, or fragments of plants that can be seen with the naked eye. Detailed examination of this evidence generally involves using some sort of magnification process that ranges from a hand lens to an electron microscope. However, when we think of forensic botany, we likely think first of macroscopic evidence. This evidence may be the most obvious application of botanical materials in forensic investigations and is the topic of this chapter.

We will also look here at cases involving the seemingly straightforward recovery of plant materials on suspects, their clothing, or their vehicles that help place them at a crime scene. Recall, for example, that part of the evidence that helped convict the serial murderer Ted Bundy included plant fragments that were caught up under his vehicle. Less obvious evidence of this sort includes the identification of plant fragments in the stomach or faeces of victims that help place them at a location even before they were murdered or link them to a suspect through a shared meal. Here, I present cases that use such data.

Plant ecology is my botanical speciality, and it can play a role in forensic botany. For example, whole assemblages of plant species can provide important evidence to locate and date clandestine graves. The plant ecology of graves is as yet the least developed area of forensic botany.

* * * * *

Jane Bock, whom we encountered in the preceding chapter, is a pioneer in the botanical forensics world, and has been pushing forward the frontiers in the field.[4] Surprisingly, Bock never expected to get involved in solving crimes. She is an emerita professor at the University of Colorado with teaching experience in botany, research expertise in desert grassland ecology and fire ecology, but no original training as an investigator of crimes.

Bock's first forensic case came in 1982, following a phone call from Denver Assistant Coroner Ben Galloway. He needed help with a case and needed someone with botanical skills. He found Bock's name in the University of Colorado course catalogue as the instructor of a plant anatomy course. One can imagine Bock grading student papers and Galloway cold-calling her to ask: 'Would you look at the stomach contents of a murder victim?'

The victim was a 21-year-old woman who had been working as an intern at a local Denver radio station. After having recently graduated from college in Massachusetts, she was staying with her aunt and uncle in the Denver suburbs. Each day, she would commute into the city. One Friday evening, she didn't return home.

Investigators found her body the next day, dumped by the side of the road a few miles away from where she worked. The coroner had already established that the stomach contents didn't seem to match what the victim was known to have eaten at lunchtime— a hamburger and milkshake with her boyfriend in a fast-food restaurant across the street from the radio station. Her boyfriend was the initial suspect, as significant others usually are. So, what did she eat for her last meal, where, and with whom? Perhaps the person she ate her last meal with was her killer or could at least provide a lead.

The pyloric sphincter muscle at the exit of the stomach closes up upon death. So, conveniently for forensic investigators, the victim's stomach retains its contents. Usually, food stays in the stomach for two to six hours, after which everything passes through to the small intestine. This information about the stomach's residence time means that what is in the stomach at death represents the victim's last meal, or at least what they ate during the last six hours of life. Investigators use a large volume of food in the stomach to indicate recency of death—a crude estimate of the postmortem interval (PMI, time since death of an individual) that is so important to determine.

In conducting the autopsy, Galloway could see what looked like salad materials in the victim's stomach, but could this be just lettuce and onions from her burger or from a later meal? If her stomach contents were from a later meal, then that would also establish that she must have been alive until early in the evening. Needing to know more, he called Jane Bock hoping that she could help with a more detailed examination of the woman's stomach contents.

Bock partnered with her colleague David Norris, an animal physiologist, to examine the stomach contents of the victim. Neither of them had forensic experience, but this first case proved to be the start of a long partnership between Bock and Norris. Galloway sent them microscope slides of the victim's stomach contents to get them started, since neither Bock nor Norris wanted to get close to the body of the victim.

We chew our food, which breaks it into small pieces, and the caustic acids of the stomach further break the food particles down, chemically dissolving the once-living tissue. The contents of our stomachs may include broken-down plant fragments (pieces of leaves and stems) along with any seeds that we may ingest (e.g. apple or watermelon seeds). Identifying these plant pieces is difficult, and back in 1982 it was not something that was routinely done as part of a victim's autopsy. 'Old-school' autopsies generally only examined what the coroner thought would be necessary (e.g. the area around an obvious bullet or knife wound). These autopsies only studied what the investigator could see with the naked eye. Pathologists performed examinations of stomach contents merely to see if the victim had a full stomach to help with time-of-death estimates.

It may seem like a daunting task to determine from a microscopic examination of a leaf fragment which species of plant has been recovered from a victim's stomach. However, among the approximately 400,000 plant species, researchers estimate that humans have used only 3,000 plant species for food. Of these, we have only domesticated 200 as food crops, and contemporary diets in North America comprise only about 70 plant food species. These 70 plant food species also include plants used in regional or speciality diets that would be informative if found in a victim's stomach contents.[5]

Even 70 possibilities need to be matched with reference samples and identified properly. A reference sample collection of partially digested stomach contents didn't exist anywhere in 1982. Bock and Norris had to improvise and act as experimental subjects themselves. So, after a trip to the grocery store for supplies, Bock chewed up samples of likely plant suspects—typical dinner vegetables. Bock chewed the samples enough times to keep anyone's grandmother happy ('chew your food 32 times'), and spat out the bolus. Norris mounted these chewed-up samples onto microscope slides, and the two investigators compared them to the samples on the slides provided by the Coroner's office.

Using this approach, Bock and Norris worked out that the last meal of the victim included red beans (Phaseolus vulgaris), cabbage (*Brassica oleracea var. capitata*), onion (*Allium* spp.), lettuce (*Lactuca sativa*), and green peppers (*Capsicum annuum*)—a salad from a Wendy's fast-food restaurant. How could they tell? The epidermal ('skin') leaf cell shapes, sizes, and features differ significantly among plant species. Onion cells are blocky, regular, and square-rectangular; lettuce cells look like a jigsaw puzzle with all cells looking different from one another; cabbage cells vary among pentagons, hexagons, and heptagons in shape; and green pepper cells are irregular and somewhat angular (Figure 6).

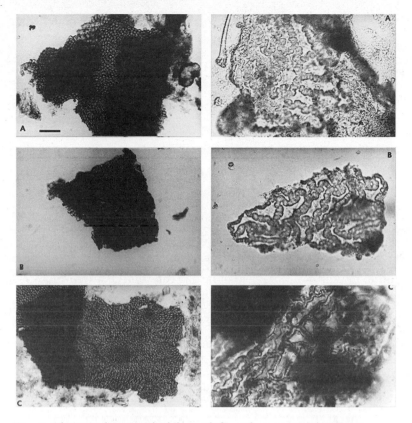

Fig. 6. Plant seed coats. Black bean (left) and pepper (right) from victim's faecal sample (top panels), victim's clothing (middle), and suspect's clothing (bottom).

With this evidence, the investigators could reconstruct a timeline of events. The victim's lunch with her boyfriend hadn't included all of these plants; therefore, it is likely that she met someone for dinner—perhaps unexpectedly, but probably a friend or acquaintance. Her boyfriend had an alibi for the evening and was eliminated from further inquiries. The evidence raised two important questions: Who did she eat her last meal with? And could that person be her killer?

Around the same time, a Texan serial killer confessed to a string of killings across the US Southwest of young women who looked similar to the victim. The prosecutors established that the victim had met him by chance at her brother's house. He had been using the phone to call about car problems. He was charming, and she didn't realize that he was a stranger to her brother. He started stalking her soon thereafter, eventually walking up to her at a bus station and starting a conversation with her like, 'Remember me? We met at your brother's house . . .'

Waitresses at a local Wendy's restaurant picked him out of a photo line-up, confirming that he had eaten at the restaurant with a young woman matching a description of the victim the evening of her murder. Much to their surprise, everything that Bock and Norris had identified from the victim's stomach was on the salad bar. Thus, setting a precedent for forensic science, Bock and Norris had linked a suspect to his victim through the identification of stomach contents.[4, 6, 7]

As Bock and Norris worked on more such cases, they developed a reference database of samples of plant fragments as they appeared in gastric contents. They built up this database experimentally; a graduate student chewed samples, which they then suspended at 37°C for 24 hours in hydrochloric acid, to mimic the conditions of a human stomach. Bock and Norris then mounted the samples on a microscope slide, and sometimes stained the slide with a dye such as safranin O, toluidine blue, or potassium iodide (Lugol's solution) to bring out certain features. For example, a drop of potassium iodide solution will stain starch grains black, which can be an important test to determine whether stomach contents contain potato (*Solanum tuberosum*), wheat (*Triticum* spp.), rice (*Oryza sativa*), lima bean (*Phaseolus lunatus*) and pinto bean (*Phaseolus vulgaris*), or pea (*Pisum sativum*) fragments.

Additionally, samples viewed under polarized light can show distinctive glow and coloration. Under polarized light, starch grains can appear as white spheres. Bock and Norris showed that plant cells retain their size and shape under these conditions, even after they have been boiled, frozen, or roasted. In 1988, the US National Institute of Justice funded the collection of images and descriptions of these samples into a reference manual.[8] This manual has descriptions and photographs, including scanning electron microscope images of 45 common food plants. Although it may not be your typical page-turner, the book was a hit within its niche genre. By comparing specimens with these reference samples, Bock and Norris could make positive identifications of the plants that the victims had eaten in many other cases that followed.

What about other bodily fluids? Bock and Norris became less squeamish about the bodily fluids that they were prepared to examine as their reputation grew. They expanded their analyses to include the investigation of plant fragments in vomit and faeces. For example, crime scene investigators sometimes find faecal materials on the bodies and clothing of victims and suspects. In one case, investigators recovered plant fragments, including black beans (*Phaseolus vulgaris*), pepper seeds, and onion leaf cells, from faecal samples of a murder victim and the clothing of her suspected killer. Nothing was unique in the samples of either the victim or the suspect. Investigators considered it unlikely that both the suspect and victim had recently had the same meals, but the suspect's clothing was stained with the victim's faecal matter. When I present this case to students, they ask me how the victim's faeces got on the suspect's clothing. I point out that relaxation of the anal sphincter is common upon death, especially during moments of great trauma. In other words, victims as they die often soil both themselves and their assailant.

Following presentation of this evidence, the suspect with his victim's faeces on his clothes was found guilty at trial.[9]

* * * * *

Perhaps examining bodily fluids, like stomach contents and faeces, is not the most obvious approach to forensic botany. More conventional approaches that immediately come to mind are when plant fragments are found on suspects or victims, linking them to each other and the crime scene. Locard's Affaire X is a fascinating early example, as are the cases linking serial killers Ted Bundy and Jamie Saffran to their crime scenes.

Could leaves of a grass less than 1 mm wide be useful forensic clues in a rape and murder case? On the morning of 16 October 1999, a groundskeeper found the body of 19-year-old Susanne Frank (not her real name) at the Emerald Beach Golf Course, an exclusive golf course in South Bahamia, Grand Bahama. Frank was a recent high school graduate living locally with her parents in the nearby town of Eight Mile Rock. She was the youngest of four girls in the family.

The crime scene was gruesome. The groundskeeper found her partially naked body lying on a fairway in the early morning as he was clearing debris following a strong tropical storm. Someone had brutally raped her and slit her throat almost decapitating her. A $1 bill had been carefully placed on her body. Nearby, police found women's clothing, a ring, and a used condom.

Police never found a murder weapon. What's worse, the storm had washed away potential hair and fingerprint evidence, DNA evidence on the condom, and other potential evidence. This lack of evidence left the police very few clues to work with.

What they knew was that Frank was last seen alive leaving a bar around midnight the night before with two men that she had met that evening. A girlfriend who had been with her thought the two men were sailors, which the police later confirmed. However, though the two sailors had gone with Frank to the bar of the Ruby Swiss Restaurant and had argued with her, they had left her there before returning to the nearby naval base, so establishing their alibi. Ken Latimer (not his real name), a local barber who Frank already knew, helped break up the argument with the sailors. Latimer and D'Andre Major (not his real name), an electrician, then joined Frank in the bar after the sailors left.

Latimer and Major were the last people known to be with her, making them the prime suspects in the case. Under questioning, they initially claimed that they had left the bar in Latimer's car and dropped Frank off at home before returning to the Ruby Swiss Restaurant to pick up some food. They did in fact go to the restaurant later that night, but that didn't exonerate them from the crime. The investigators found no evidence in Latimer's car, but obtained clothing and shoes from him and Major. Major's clothes yielded no evidence as his girlfriend had washed them, although she later testified to finding them covered in grass.

Criminalist Elbert Ferguson examined Latimer's clothes and shoes and found grass fragments on his shoes and socks. Not surprisingly, he couldn't identify the fragments because he wasn't a botanist—but he knew one. He had once taken a class with Jane Bock, so he sent her the specimens to identify.

There are more than 10,000 species of grasses worldwide, and dozens in the Bahamas, but Bock was able to identify the fragments as leaves of Bermuda grass (*Cynodon dactylon*)—a commonly planted grass. But the extreme narrowness of the leaves showed that this Bermuda grass was the special 'almond variety'—a fine-leaved strain developed by the groundskeeper unique to that golf course. These fragments matched samples collected from the golf course and found on the body of the murder victim, providing a link between the victim, the suspect, and the crime scene—a perfect example of the four-way linkage theory (Figure 5).

Confronted with this evidence, Latimer and Major accused one another of the crime. Latimer confessed that he had been at the scene as Frank jumped out of Latimer's car and ran across the golf course trying to escape the pursuing Major. Tellingly, though, there were no grass fragments on Latimer's shorts or shirt, as there would have been if he'd lain down on the ground to take part in the rape and murder. This evidence supported the view that Latimer had been at the scene but had not taken part in the rape and murder.

Both men were tried. After Latimer confessed to being a witness to the crime, he received six years' imprisonment. The court convicted Major of rape and murder, and the judge sentenced him to death. He was later resentenced following appeal to manslaughter and 10 years, following a successful argument by his defence that he was under the influence of alcohol at the time of the crime and had no 'intent to kill'.[10, 11]

* * * * *

What if a case isn't homicide but an accident or even suicide? Plants can help make this determination. A young woman's body was found lying in the gutter of an alley in the Taiwanese capital,

Taipei. Video evidence from surveillance tapes showed a truck passing through the alley but was inconclusive. Was this a hit-and-run? Did the perpetrator move the body into the gutter to hide the accident? An ambulance took the woman's body to the local hospital, where investigators found a berry and a stem fragment in her hair that they tentatively identified as from plants in the Solanaceae family.

While searching the scene where the body was found, investigators found part of a broken stem in the gutter. The investigators looked up and could see potted plants on the edge of a railing 3.5 m above the gutter. They identified the pot plants as blackberry nightshade (*Solanum nigrum*), a plant native to Eurasia. While the unripe berries of blackberry nightshade can be toxic, the cooked ripe berries and leaves of blackberry nightshade are used as a food and in traditional medicine. The fragments in the gutter and in the woman's hair matched the broken stems of the potted plants. Although these were herbaceous plants and not woody, the wind was not strong enough to break the stems of the potted plants. The railing was too high to be reached from the ground. At this point, the investigators started to think the woman had committed suicide instead of having been a hit-and-run victim.

The autopsy showed the woman's injuries to be consistent with a fall from the top of the building. The investigators inferred that her head and hair had contacted plants on the way down. Furthermore, the woman had a known history of depression and previous suicide attempts. Thanks to the identification of the plant fragments, they ruled her untimely death a suicide and not a homicide.[12]

Sometimes plant evidence helps to determine the cause of accidents,[13] even those that occur when a crime is being committed. In one case, a woman brought her 28-year-old Romanian husband to a hospital in Italy with electrical burns.[14] He died a few hours later. His wife told investigators that he had been repairing a chandelier at home when he received an electrical shock and fell off the ladder he was using. Investigators were suspicious, so they checked things out.

Indeed, the investigators found a ladder on the ground under a chandelier in the man's two-floor apartment. They also found electrical wires on the chandelier and blood splatters on the ladder and floor. During the autopsy, the forensic pathologist found that the man's body had severe burns on his arms, chest, and left leg and cutaneous necrosis consistent with burns from high electric voltages. The pathologist also found head trauma, brain haemorrhage, and fractures at the base of the skull consistent with a fall from some height. However, the evidence aroused suspicions because burns from low-voltage residential electrical shocks don't cause such deep dermal damage as that observed on the body. Most surprisingly, investigators found plant fragments and fruits in the hair, body, and on the clothing of the victim. Botanists from the nearby University of L'Aquila identified the plant fruits as Spiny cocklebur (*Xanthium spinosum*). This plant is an annual weed of arid, uncultivated, rural areas, with yellowish, egg-shaped, two-seeded fruits covered in hooked spines that are easily caught up on the clothing of people and fur of animals.

The electric company had a record of a local power cut-off signal at a junction pinpointed to an area outside the city where there was an electric pylon. This site became a secondary crime scene. Plants of *X. spinosum* occurred at this site. Piecing everything together, investigators established that the victim was electrocuted while trying to steal copper wiring from the cable on the electric pylon. The shock caused him to fall to the ground into vegetation where X. *spinosum* was growing, and fruits of the plant became attached to his body, hair, and clothing. The man's wife carried his body back to their apartment and set up the primary—and false—crime scene. The botanical evidence in this case was critical in linking the victim to the correct crime scene and in revealing the false account by his wife. The accident had occurred, but not at home, and it was actually an unfortunate consequence of committing a crime.

* * * * * *

Determining the time of death of a victim, the PMI, is one of the most important things that a crime scene investigator needs to do. Time estimates derived from plant phenology (when a plant grows), plant growth rates (how much a plant grows over a period of time), or plant senescence (the changes associated with the process of aging leading to death) can help.

Finding a single human bone on top of vegetation helped provide a phenology-based estimate of the PMI in the Gold Head Branch murder.[6] Hikers found a skeleton deep in a ravine in Gold Head Branch State Park, 50 miles south-west of Jacksonville, Florida. Investigators needed to know how long the skeletal remains had been there to help them identify the victim and learn the cause of death, so they enlisted the help of forensic botanist David Hall.

By the time that Hall reached the scene, crime scene technicians had removed the bones. Nothing remained but an outline of the body in the undergrowth and many technicians' footprints. To Hall's distress, any forensic evidence had been trampled on and destroyed, and indications of vegetation that might have grown in and around the skeleton were gone. Hall needed to examine an undisturbed, *in situ* bone. Fortunately, examination of the skeleton showed two missing bones: a jaw and a leg bone. They found the jawbone and the long tibia from the leg after a careful search in which more than a dozen officers walked at arm's length through the ravine with instructions not to pick anything up. Hall returned to the scene a second time with more hope. The jawbone was of no use as it was lying on open sand. The tibia, however, was more promising.

The tibia bone was lying across a 5-inch-tall *Quercus laevis* (turkey oak) seedling, bending it to the ground. *Quercus laevis* is an unremarkable shrubby oak common in scrub habitats of the south-eastern United Sates. Its leaves are deeply and narrowly lobed, resembling a turkey's foot, hence the common name of turkey oak. There was a black band across a mature leaf under the bone where the green chlorophyll had degraded in the dark shade. Hall estimated that the leaf must have been under the bone for two weeks to kill all the chlorophyll. However, the exact time depended upon when fully mature leaves of *Q. laevis* had developed that year: the phenology of the plant.

After speaking with local park rangers and maintenance workers, Hall learned that spring had been late, with the deciduous *Q. laevis* coming into full leaf in the first or second week of March. Hall added three weeks for the leaves to mature and two weeks for the chlorophyll to degrade to arrive at an estimate that the leg bone had landed on top of the turkey oak leaf in mid to late April. The bone had gnaw marks from a dog which may have dragged it from the decomposing body. This evidence of gnawing on the bone is not surprising as scavengers commonly disturb human cadavers, making CSIs race to beat them to it. Based on his knowledge of human body decomposition, and being careful not to overstep his area of expertise, Hall estimated that the body must have been five to six months old to be sufficiently decomposed to allow a dog to pull or drag away part of the leg. The time of death must have been the end of September. Hall's estimate allowed investigators to identify the victim from missing person records.

Despite the PMI estimate, it took 20 more years to solve the crime. Two decades after the gruesome murder, a witness finally came forward to identify the killer and confirm an early October, execution-style shooting. Hall's estimate based upon the condition of a single leaf under a single bone was remarkably accurate—within a week or two of the timeline described by the witness.

At the time, Hall was a botanist with the University of Florida herbarium. As knowledge of his expertise spread among the crime investigation community, he was offered and took on so many cases like this that he started his own forensic consultancy. We have already come across him in his role as an expert witness in a child murder case, in which prosecutors pitted his testimony against Jane Bock's for the defence.

How did Hall come up with the estimate of the time taken for chlorophyll degradation to occur in the leaf covered by the bone? Gut feel and 'expert knowledge' is not a valid answer, especially as courtroom rules for expert witnesses allow the validity of scientific methodology to be challenged. Hall has conducted experiments covering and uncovering leaves at different intervals and noting the changes in coloration. Several samples of leaves will be buried or covered, ideally at crime scenes. These experiments need to be conducted under the same conditions at the scene and as soon as possible after the plant material is discovered.[15]

In the Gold Head Branch murder case, determining the PMI was estimated based on plant phenology. Measuring the growth rate of plants is a different but alternative approach. Even the growth rate of a moss can help. Such was the case when police investigated the skeletonized remains of a man found in a wooded area of northern Portugal in February 2008.[16] Clothing and the remains were consistent with a 61-year-old man who had gone missing six years previously. The man was an indigent with no known relatives and no clinical or dental records. Nevertheless, a PMI was needed.

The remains were mostly skeletonized and disarticulated, with only some soft tissue desiccated in one foot inside the remains of a sock or boot. Plant roots were growing through the remains, especially the ribcage. Investigators identified the roots as belonging to shrubs in the rockrose (Cistaceae) plant family. Examination of growth rings in the roots gave a PMI of three years. Green algae and mosses were growing over some of the bones and clothing. In particular, they identified four mosses as *Bryum capillare, Campylopus flexuosus, C. introflexus,* and *Hypnum cupressiforme.*

The mosses found colonizing the bones in this case grow through a monopodial growth pattern, in which the apical meristem (the area of dividing cells at a shoot tip) continues growth over more than one growing season. Other mosses have sympodial growth, where the apex ceases growth and new growth comes from side branches. The length of the successively produced side shoots in monopodial mosses declines through the growing season, so botanists can see the start of a new season's main-stem growth from where the production of new longer side shoots starts. Botanists refer to the growth of the main monopodial stem in a season as a growth segment. A count of the number of growth segments provides an estimate of how long the moss has been growing over the bones. In this case, PMI estimates from

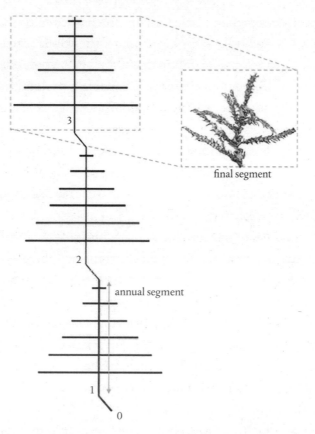

final segment

annual segment

Fig. 7. Diagram of the main stem of a moss illustrating the growth process (monopodial growth). 0 represents the newest bud and years 1, 2, and 3 represent the annual segments.

the mosses colonizing the clothes and bones were one to three years (Figure 7).

There are, of course, limitations to using mosses to estimate PMI. The resulting estimate is a minimum as mosses apparently do not colonize human remains until the soft tissues have decomposed and the bones are skeletonized, and the time that takes varies depending upon environmental conditions. Moss colonization may take even more time if the skeletonized bones are inaccessible to colonization due to being covered, buried, or submerged. And not all mosses are suitable for estimating PMI, particularly those that exhibit sympodial growth. Some monopodial mosses even show two annual growth pulses, which could lead to an overestimation of the PMI, while some aquatic mosses show continuous growth. Accurate species identification and knowledge of their growth are essential for the precise application of bryophyte forensics.

An experiment on plant senescence in the early 1990s provides an example of a case that has achieved some notoriety—the case of the wilted sunflowers.[6, 7, 17] In fact, there is a made-for-TV drama film *Ultimate Deception* (1999, also known as *Ultimate Betrayal*) based upon it. The names of the principal characters in the film and the location were changed, but the general story is close to what happened. I describe the actual case here.

In July 1991, a highway maintenance worker found the body of a woman later identified as Terra Ikard in a ditch 30 miles outside Aurora in Arapaho County, Colorado. Her killer had shot her twice in the chest and once in the head. Unfortunately, the police did not find a weapon at the crime scene. The killer had dumped Ikard's body in the ditch and had covered the corpse with uprooted sunflowers (*Helianthus annuus*) that had been growing in an adjacent field. Her body was bloated and decomposing, showing the early stages of decay. Investigators needed a PMI estimate. Crime technician Jack Swanburg noted that the sunflowers were wilted, but not so decomposed as the body. He wondered if the sunflowers could help determine the PMI, so he called Jane Bock.

Realizing that she needed to determine how long it took for sunflowers to wilt, Bock told Swanburg to pick some fresh sunflowers from the adjacent field and bring them with the wilted sunflowers to her laboratory at the university in Boulder, Colorado. Bock then bagged and froze the wilted sunflowers; these would become the comparators in the simple experiment that she would conduct.

She placed the recently picked bunch of sunflowers out in the University of Colorado rooftop greenhouse and started a series of daily observations, carefully noting how they changed as they wilted. After seven days, the sunflowers in the greenhouse resembled the condition of the wilted sunflowers when they were recovered from on the body; after six more days, they suddenly shrivelled. Based on this simple experiment, Bock gave the police a one-to-two-week PMI estimate. Her estimate matched the PMI estimate made independently by entomologist Boris Kondratieff from the developmental stage of maggots collected from the corpse.

Bock's botanically derived PMI estimate from the wilted sunflowers helped establish the timeline of events surrounding the death and dumping of Ikard's body, and was an important part of this unusual case. The 'sunflower killer' Ralph Takemire was convicted and sentenced to life without parole. He died an inmate in the Colorado prison system on 6 March 2006.

* * * * *

Serial killer Herb Baumeister buried several of his victims on his 18-acre Fox Hollow Farm in Westfield, Indiana. Similarly, Canadian landscaper and serial killer Bruce McArthur buried his victims in garden planters. Other serial killers like Ted Bundy established 'body dumps', such as his Taylor Mountain site where he left decapitated heads of his victims, revisiting the site several times.

Finding body dumps of serial killers is hard enough, but is made tougher still when the bodies are actually buried: of course, some killers bury their victims' bodies rather than simply dumping them, hoping the bodies will not be found and their crime will remain undetected. Indeed, investigators never find the bodies of many serial killer victims for this reason. Regardless, many serial killers return to their body dumps and the clandestine graves of their victims. Criminologists think these killers return to the crime scene to help them feel a sense of control over their victims, relieve in part their burden of guilt, or re-live in their minds the events related to the murders.

Finding body dump sites is a challenge that is poorly understood. Fortunately, botanical evidence can help with this challenge. A patch of ground bare of vegetation can indicate an area where a body has decomposed on the soil surface. As decomposition of a body occurs, microbes and insects break down the bodily chemicals. The decomposing body releases caustic fluids that will kill existing plants and prevent other plants from growing.

Decomposition follows a general series of events that vary depending on environmental conditions and whether the body is on the soil surface or buried. Bodies placed on the ground decompose faster than those that are buried, and the skeletal remains can become scattered by scavenging animals or dispersed by flowing water. Decomposition rates are faster in moist and warm conditions compared with dry and cold, or dry and windy conditions. High rates of bacterial, fungal, and insect activity hasten chemical breakdown in moist, warm conditions. By contrast, in dry and windy conditions, exposed bodies can desiccate with very little decomposition. Identification of bacteria, fungi, and insects growing in or on bodies can help provide a PMI estimate.

A decomposing body adds about 2.6 kg of nitrogen to the soil; that's about 50 times the recommended rate for garden fertilizer. As enhanced levels of nitrogen become available in the soil, the soil microbes comprising the 'necrobiome' are stimulated. A halo of lush plant growth may later surround the bare area around a decomposing cadaver, known as the cadaver decomposition island (CDI). This 'greening effect' may increase leaf chlorophyll production in plants. Researchers speculate that the resulting altered spectral signal of plants growing around a decomposing body can be picked up by the sensors on a drone, allowing remote sensing to find cadavers.[18]

As the cadaver dries out, pioneer weed plants may colonize the CDI because of the high levels of available soil resources and low levels of competition from other plants. The remains of the cadaver will become a scattering of disassembled skeletal bones, or articulated remains held together with remaining ligaments, amidst a distinct island of vegetation.[19]

By contrast, soil disturbance from digging a grave and burying a body will lead to loose, water-retentive soil causing a flush of germinating seeds in the soil and a vigorous growth of weedy plants. Ecologists are very familiar with the concept of the soil seed bank, where seeds remain viable, often for decades, until changes in temperature, light, or water, or all three factors stimulate germination. As the soil settles above the decomposing body, a depression in the ground will occur. Initially, the vegetation is suppressed, but with time the vegetation growth on the burial surface may be lush, perhaps with differences in flowering and fruiting patterns compared with the plants in the surrounding vegetation, and perhaps with distinctive species.

Forensic botanists need knowledge of the local flora to make these comparisons. These are all signs the investigators can use along with other methods such as ground-penetrating radar to locate clandestine graves. Of course, the older the burial, the more likely that vegetation on the disturbance will resemble the surrounding plant communities, rendering the grave indistinguishable from a botanical perspective.[20]

A particularly novel example of using a plant to identify a clandestine grave was the occurrence of a fig tree (*Ficus carica*) growing through the opening to an underground cave on a beach in Cypress. The fig tree was thought to have germinated from seeds in the stomach of a body that had been dumped in the cave. Fig trees do not naturally occur in that area and a sharp-eyed local man became suspicious when he spotted the tree. Investigators subsequently found the bodies of three Turkish resistance fighters in the cave who had been kidnapped and disappeared 40 years earlier. One of the men was reported to have been eating figs just before he was kidnapped, lending credence to the proposed origin of the fig tree. As you might expect, the tabloid media liked this story, and it circulated widely under headlines such as 'Murdered man's body found after tree unusual for area grew from seed in his stomach' in the UK's *Daily Mirror*.[21] Other accounts suggest that the roots of the fig tree, while in the cave, were some distance from the bodies.[22]

Investigators found the diminutive annual plant *Teesdalia nudicaulis* (shepherd's cress) and other acidophilic (acid soil loving) plants to be potential indicators of the disturbed-ground characteristic of clandestine graves in an experiment using pig cadavers in Italy. Most research that is focused on the ecological and botanical features associated with clandestine graves has been

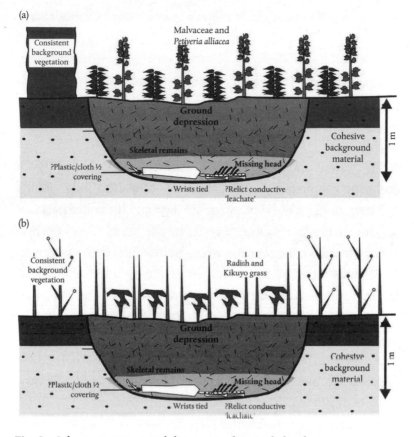

Fig. 8. Schematic annotated diagrams of typical clandestine grave encountered in Latin America in (a) tropical rainforest and (b) semi-rural environment.

conducted on human surrogates such as pigs. These studies have documented the succession of plants that colonize a grave, showing the abundance of certain indicator species or types of plants (e.g. weed annuals early on; see Figure 8).[23, 24, 25] Investigators need to conduct comparable studies on human cadavers, but these are understandably difficult to conduct as an appropriately replicated statistical experiment due to limitations on cadaver availability and time.

Nevertheless, 'body farms', as they are euphemistically called, can help address the research needs associated with dumped bodies and clandestine graves. These are specialized decomposition facilities where human cadavers are studied. Dr William Bass established the first of these facilities in 1987 at the University of Tennessee, Knoxville, Forensic Anthropology Center (FAC), following enquiries by the local police about how long victims had been in place at crime scenes.[26] Bass and his colleagues laid out donated cadavers either directly on the ground (in cages and behind fences to keep out scavengers) or buried. At intervals, researchers recorded the condition of the cadavers and related decomposition to environmental conditions.

Today, there are at least seven such decomposition facilities in the United States, as well as the Australian Facility for Taphonomic Experimental Research (AFTER), which opened in 2016, and the Canadian Secure Site for Research in Thanatology (the scientific study of death) which opened in 2019. And there are proposals for such facilities to be established elsewhere across the globe including India, the Netherlands, and the UK. Research at these facilities helps anthropologists and other specialists to better understand the chemical, microbial, and entomological stages associated with cadaver decomposition.

My institution of Southern Illinois University Carbondale is home to the Complex for Forensic Anthropology Research facility (CFAR) run by my colleague Gretchen Dabbs. CFAR accepted its first donated body in 2012. Among other insights, Dabbs's research has shown that various fungi colonize the desiccating skin on cadavers exposed to the drying winds in southern

Illinois. This observation of fungal colonization has the potential to help with establishing a PMI of dumped bodies found under similar environmental conditions, but much more research is necessary.[27]

* * * * *

Sometimes it is not so much the identity of plant fragments that can help solve a crime, but the known ecology or natural history of the plants identified as evidence. For example, one search, for the remains of a missing hiker, presented a particularly challenging botanical case and involved members of a group of volunteer forensic experts, NecroSearch International (NSI), which helps law-enforcement agencies to search for and recover human remains.[28] It took several years, but plant fragments caught up in a recovered scalp helped the group locate the rest of murder victim Michele Wallace's remains. In 1974, lone hiker Wallace went missing in the Gunnison area of the Colorado Rockies. She was last seen alive giving local sheep rancher Roy Melanson a ride in her car on 27 August 1974. A week later, her dog Okie showed up after a rancher protecting his stock shot it. Roy Melanson was a known serial rapist and became a suspect early on, especially since he was found in possession of her car and camera. However, without a body, there was not enough evidence for a conviction.

In 1979, a hiker was shocked to find a mass of human hair along a logging road high on the Kebler Pass 10,000 feet above sea level. The hair was parted into two braids, just like they were in the last known photo of Wallace found on her camera, which Melanson had pawned. The hair matched strands found on a brush belonging to Wallace. Despite this evidence, the case remained a cold case.

Finally, in 1990, police investigators asked NSI member and botanist Victoria Trammel to examine the hair. She observed that the exposed hair on the braids was sun-bleached, whereas the hair was dark brown on the inside. This difference in coloration likely indicated that the body had not been buried. Plant fragments caught in the hair included needles of subalpine fir (*Abies lasiocarpa*), one needle of Engelmann spruce (*Picea engelmannii*), a stem fragment from the shrub mountain lover (*Paxistima myrsinites*), and non-conifer wood fragments thought to be of aspen (*Populus tremuloides*). Based on the ecology of these species, Trammel concluded that the hair had spent time exposed on a cool, moist, north-facing slope in a mountainous habitat above 9,000 feet that had not been recently exposed to fire.

An NSI team, comprising an anthropologist, archaeologist, and a naturalist, joined police investigators on a systematic, hands-and-knees 'pedestrian sweep' search of the area downslope of the logging road. They first found Wallace's skull, then her skeletonized remains and clothing fragments. Her remains had been scattered by wildlife, moving water, or perhaps even wind that can create enough force to move skeletal elements. Dental records confirmed her identification. This is an excellent example of a case in which ecological knowledge of habitat characteristics of the plant leaves and fragments allowed investigators to narrow down the search area sufficiently to find the victim's remains. Melanson was tried and the plant evidence helped secure a conviction of first-degree murder on 1 September 1993.[28, 29]

* * * * *

Each example in this chapter has illustrated a very different way in which macroscopic plant remains and fragments can be of value in forensic investigations. In the next chapter, we will look at the use of microscopic plant, algal, and fungal evidence from pollen and spores. The identification of these microscopic structures requires a level of expertise that most conventionally trained botanists do not have. While the perpetrator of a crime may well be unaware that plant fragments from a crime scene are caught up on their clothes, they have virtually no way of knowing that dust-like structures may hide in the mud on the soles of their shoes.

4

EVERY PARTICLE TELLS
A STORY

> Nearly every manufactured ancient article, has trapped
> somewhere inside it, pollen from the age and location of
> its manufacture . . . every particle tells a story.
>
> Clive Langmead (1995) *A Passion for Plants*[1]

The abduction and brutal murder of two 10-year-old school-girls, best friends, shocked everyone in the United Kingdom in 2002. Pollen helped crack this much-publicized case. The two girls had been attending a family barbecue on the afternoon of 4 August when they sneaked out to buy sweets from a local shop. They never returned.[2, 3, 4]

After a massive search, police found the matching Manchester United football jerseys that the girls had been wearing to the barbeque. Someone, presumably the girls' kidnapper, had burned and dumped the two jerseys in a stockroom rubbish bin in their local school. On 17 August, a gamekeeper found the partially burnt bodies of the two girls in a roadside ditch a few miles away from the town of Soham, where the killer had abducted them. By this time, the authorities had a suspect, a known paedophile who was working as a caretaker at Soham Village College, the school attended by the girls. How a known paedophile landed a job at a school in the first place is a worrying aside. Nevertheless, he lured the girls into his house on the pretext that his partner, one of their favourite teachers, was there to see them. In fact, the teacher was away for the weekend, so she did not take part in the murders, but the authorities later convicted her for assisting the perpetrator in covering up the crime.

A key part of the evidence implicating the suspect in the murders was the pollen and soil evidence that linked his car to the body dump site. A forensic botanist found that pollen in soil on the chassis, spare wheel, and footwells of the suspect's car and on his shoes matched the vegetation in the ditch. This pollen, along with evidence of the path the killer took to the ditch from the pattern of regrowth of the nettle (*Urtica dioica*) plants he had trampled, helped secure a conviction of the suspect.

Regardless of what we may see in sci-fi movies, only on rare occasions are microscopic plant (pollen), diatom (types of single-celled algae), or fungal (spores, hyphae, or fruiting bodies) structures in high enough abundance that we can see them as anything more than 'dust'. Yet just like plant leaves, wood, and other fragments discussed in the previous chapter, these organisms can provide key evidence in criminal and civil cases.

* * * * *

A pollen grain is a remarkable, armoured structure that carries the male gametophyte of a seed-producing plant (i.e. an angiosperm or a gymnosperm) from an anther on a stamen of one flower to the stigma, the top of the stalk-like style in the pistil of the same or another flower. Sometimes, the transfer of pollen is from the anthers to the stigma in the same flower or in different flowers on the same plant. In both of these cases, this transfer allows for self-fertilization. By contrast, in some species, the stamens and stigmas are in flowers on different plants of the same species, ensuring cross-fertilization. The armour on a pollen grain is the remarkably decay-resistant and chemically complex polymer sporopollenin, which occurs in the outer part (exine) of the pollen wall. When a pollen grain lands on a compatible stigma, it germinates, sending a pollen tube growing down through the style to the ovary at the base of the pistil to enact fertilization.

The dispersal of pollen from anther to stigma can be by one of several carriers or vectors, including wind, water, gravity, and organisms such as insects, beetles, bats, or birds. The shape and structure of a pollen grain provides a diagnostic clue of how it is dispersed. Wind-dispersed pollen, such as that of grasses and many trees, is fairly smooth and spherical. Some pollen grains have air bladders that help them stay aloft in wind currents. The two air bladders on pine pollen (*Pinus* spp.) give the pollen grain the appearance of a Mickey Mouse head. By contrast, the walls of insect-dispersed pollen are uniquely shaped, frequently exquisitely sculpted, and often ornamented with an array of hooks and spines that allow them to hitch-hike on insect body parts (Plate 2).

These differences in morphology allow the pollen from different plants to be distinguished, usually down to the genus level. In other words, researchers can distinguish pollen from oaks in the genus *Quercus* from pollen from beech trees in the genus *Fagus*. Ferns, fern-like plants, and bryophytes produce not pollen but spores that are also involved in reproduction and are similarly tough and resistant to decay. Their spores are wind- or water-dispersed and have few distinguishing features.

Palynology is the study of pollen. Forensic ecologist and palynologist Patricia Wiltshire pioneered the 'picture of place' approach, which involves obtaining a description of the vegetation at crime scenes from pollen and fungal spore samples recovered from suspects, vehicles, and victims, among other surfaces. Collectively, these microscopic samples are known as palynomorphs.

Wiltshire compares the list and abundance of species of pollen collected from evidence with samples collected from the soils, sediments, and vegetation associated with a suspected or known crime scene. Forensic scientists do not choose these latter 'comparator' samples randomly, like control samples in a scientific survey or experiment. Rather, comparator samples are chosen deliberately to represent the area. Investigators must take enough samples to allow for the extreme variability through time and across space in deposition rates, known as the 'pollen rain' of pollen and spores. In this approach, an investigator should sample any surface with which a victim or suspect may have been in contact. Pollen accumulates in the body's mucous membranes (e.g. up our noses) and in all sorts of nooks and crannies associated with clothing. In vehicles, pollen will be in the mud on car tyres, in the radiator grille, under the chassis, and down in the seat covers.[5, 6]

* * * * *

Palynology played a critical role in the evidence gathered to prosecute war criminals in The Hague following the Bosnian wars of 1995 in the former Yugoslavia. The discovery of mass graves bore testimony to the horrific massacre of civilians in Srebrenica by units of the Bosnian Serb army. From 1997 to 2002, the United Nations International Criminal Tribunal for the former Yugoslavia (ICTY) exhumed these mass graves as part of data gathering and environmental profiling. Serb paramilitary groups had exhumed the original, primary mass graves and transported the bodies to several secondary grave sites, making identifying the victims more difficult (Figure 9).

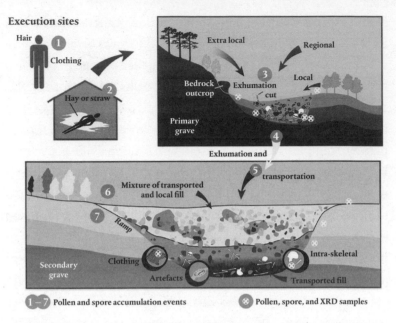

Fig. 9. The sampling contexts of the environmental investigations for the Bosnian war crimes.

ICTY investigators gathered 240 samples from 24 sites and analysed 65 pollen subsamples. Investigators took samples from bodies, clothing, and grave fill close to and far from body parts at primary and secondary grave sites. After processing and examining the samples, they compared pollen and spore profiles, summarizing the percentages of different pollen and spores from the primary and secondary grave sites. The bodies had pollen and spores from where the people lived, where they were first buried, and where they were secondarily buried.

The dominant taxa in the pollen and spore lists from the primary sites were quite different to each other. But there were similarities in the pollen taxa found between primary and secondary sites. For example, the Lazete primary site had pollen of cereal (mostly maize, *Zea mays*), grass, pine (*Pinus*), and spruce (*Picea*) that was also found in samples from the Hodzici Road secondary site. This similarity in the pollen profiles meant that presence of these taxa on samples from the secondary site provided links to the particular primary site at which the Bosnian Serb army had originally buried the bodies. Investigators confirmed these findings and strengthened them by additional mineralogical evidence from the sediments, clothing, personal effects, ballistics, and documents. This comprehensive environmental profiling approach was novel and set a new precedent in war crime investigations.[7]

In a remarkable civil case, palaeoecologist Rolf Mathewes from Simon Fraser University, British Columbia, found microscopic pollen that supported a land claim based on the oral history of the local Indigenous People. This novel forensic palynology case included the evidence presented as part of an indigenous land claim in Canada in *Delgamuukw vs The Queen*.[8, 9] Following a protracted, decade long case, the Canadian Supreme Court finally affirmed in December 1997 the legitimacy of oral histories (known as *adaawk*), traditional laws, and continuing governance of lands where Indigenous Peoples have never signed a treaty with the Crown.

The oral history passed on by the Gitxsan and Wet'suwet'en Nations' Chief Delgamuukw recounts, substantiates, and validates the continuous presence of the Indigenous Peoples in a large area of British Columbia, Canada. In this case, the oral history includes the account of a massive debris flow coming down from what we now call the Chicago Creek drainage as it flows into the Skeena river. This debris flow followed the breach of the upstream lakeshore and knocked down and uprooted the streamside forest. The flow of debris also dammed Seeley Lake, near the present-day village of Hazelton, which had drained into the Chicago Creek. Delgamuukw represents this event in oral history as the appearance of the Medeek, a giant, supernatural grizzly bear appearing out of the river. The oral history recounts that the Indigenous Peoples followed the path of destruction up the mountainside, where they found that the devastation disappeared into the Lake of the Summer Pavilions. They concluded that the bear must have appeared out of the lake.

Palynological evidence supported the occurrence of this prehistoric event. Rolf Mathewes extracted sediment cores from Seeley Lake for radiocarbon dating and examination of pollen and plant macrofossils. He found that the pollen and macrofossils recovered from the cores showed a rapid increase in abundance of aquatic plants, including pondweeds (*Potamageton* spp.) and coontail (*Ceratophyllum* spp.). This change in the vegetation reflected a rise in the water level of Seeley Lake after its outlet became dammed by the debris coming down the mountainside. There was also an increase in pollen of green alder (*Alnus crispa*; now *A. alnobetula* subsp. *crispa*) from colonization on the debris along the slopes of the lake. Mathewes established through radiocarbon dating of wood within the slide debris that the event occurred 3,500 ± 150 years ago.

The pollen examined by Mathewes confirmed the accuracy of the oral history over a time period of three millennia, leading to the court finding for the plaintiffs. Under the precedent-setting ruling from the Supreme Court, Canadian provinces cannot extinguish Indigenous Peoples' titles to land, and they must be consulted and compensated with regard to any development on these lands. This was a momentous legal decision based on forensic palynology.

*****

Diatoms are a group of unicellular, mostly aquatic algae that can be useful in forensic investigations. There are approximately 100,000 species of diatoms. They are truly things of beauty (Plate 3). The presence of fucoxanthin, an accessory pigment, lends them a golden-brown colour. A silica-based (glass-like) cell wall or frustrule encases the living diatom cell. This frustrule is made of two halves that fit together like two halves of a vintage hat box, and is highly ornamented with species-specific pores and grooves. Most diatoms are bilaterally symmetric, known as pennate, while a few are radially symmetric, known as centric. The acid and decay-resistant frustrules are readily recovered from samples even after the actual cell has died. Some diatoms are terrestrial in soils, but most are aquatic in fresh- or saltwater.

Diatoms are highly abundant, with their populations varying seasonally in lakes, rivers, and ponds. Peak blooms occur in the spring and autumn. Diatoms can provide excellent forensic evidence because, although there are many cosmopolitan diatom species, the relative abundance of different species is diagnostic for a particular location and/or time of year.

The group of teenage bullies regarded themselves as having fun when they attacked two younger boys with knives as they were fishing by a pond in Connecticut, New England. They then bound the two boys up with duct tape, beat them with a baseball bat, and dragged them into the pond to drown before absconding with the boys' bicycles. Fortunately, one boy freed himself, saved his friend, and went for help. Police later arrested the three suspects, who they noticed were wearing muddy sneakers.

Peter Siver from the Department of Botany at Connecticut College examined the sediments on their sneakers and compared it to sediments on the victims' sneakers and to pond sediment. Siver found 14 of 25 species of diatoms in common among the samples, with no statistical difference in their relative abundance. And he found the same ratio of three freshwater *Eunotia* diatom species in all the samples being compared. A pond or lake, and not a stream or river, was indicated by occurrence of the scaled chrysophyte *Mallomonas caudata* as the dominant planktonic algae in all samples. The diatoms and the planktonic algae provided compelling evidence that all the sediment samples were from the same locality and placed the suspects at the crime scene. Presented with this evidence, the suspects pleaded guilty to a range of felony charges and were incarcerated.[10]

In a classic 'But I wasn't there' case, Nigel Cameron, a diatom specialist of University College, London, used diatoms to link a suspect to a crime scene. A member of the public found a woman's body in a river in south-west England.[11] The husband was the obvious suspect, but he denied his presence at the crime scene. The police asked Cameron to investigate after police surveillance caught the suspect trying to dispose of some of his clothing. The suspect was right to be concerned about his clothing, as diatoms are readily caught up and retained on fabric that is immersed in water. Indeed, the 'persistence dynamics' of diatoms and other trace evidence on clothing is an active area of forensic research.[12, 13]

Cameron used hydrogen peroxide to prepare diatom samples from pieces of the recovered clothing—peroxide dissolves away just about everything except any diatom frustules. He then compared the diatoms found in these samples with the diatoms from water samples he collected upstream, downstream, and at the location where the woman's body was found. The stretch of river where the body was found provided the best match to the diatoms on the suspect's clothing. The husband's alibi was broken.

The diatom evidence placed the accused husband in the water at the point where the body was found. The English Crown Prosecution Service secured a conviction and the judge convicted the husband, sentencing him to life in prison for his wife's murder.

Perhaps the most fascinating use of microscopic evidence is the potential of diatoms to determine whether a body found in water had drowned or been killed earlier. When investigators pulled a 58-year-old woman's body out of the Hudson River, New York, in May 1996, one of the first questions that they asked was, 'Did she drown, or was she already dead when her body was dumped into the river?'

In the 'Murder in the Hudson River' case, diatoms recovered from the dead woman's femoral bone marrow were consistent with the victim drowning in the river. A person who is drowning gulps in large quantities of water with such force that in the moments before death, there are ruptures of the alveolar-capillary membranes in the lungs. Small structures such as diatoms pass through the blood–lung membranes into the bloodstream and become transported by the still-beating heart of the drowning victim around their body and into the bone marrow. After death by drowning, specialists can then find diatom frustrules in samples taken from the femoral bone marrow (Figure 10). A person already dead who is dumped in water will have their lungs filled with water, but there will not be any diatoms in their bone marrow.

Investigators have used the 'diatom test' widely since its development in the 1960s by Belgian investigators led by J. Timperman.[14] Moreover, as we saw earlier, a comparison of the species

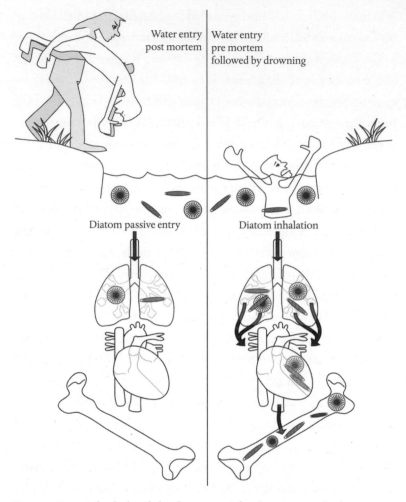

Fig. 10. Principles behind the diatom test for drowning, which compares a live drowning (pre- or ante-mortem immersion) to the dumping of a dead body (post-mortem immersion).

of diatoms in the water in the lungs with those in the water body itself can help narrow down the location of where a body was dumped or drowned.

In the Hudson River case, the obvious suspect was the woman's 63-year-old live-in boyfriend. His nephew was an accomplice and testified in exchange for immunity. He said that the boyfriend had initially drugged her with diphenhydramine, which forensic investigators found in the bloodstream of her corpse. However, she regained consciousness when he dumped her body into the river, at which point the boyfriend strangled her underwater.

Neck wounds on the victim were consistent with strangulation. The presence of diatom frustules in her femoral bones was consistent with drowning during the strangulation. And the species of diatoms in her bone marrow were the same as in river water sampled from the location where investigators had found her body. Notably, they included the diatoms *Diploneis* and *Nitzschia/Hantzchia*. They found the same diatoms on the accused's shoes, watch, and wallet after investigators recovered them, abandoned close to the river bank. This concordance of diatoms was consistent with death in that location and not elsewhere, and they linked the accused to the crime scene. Presented with this evidence at the accused's trial, jurors found him guilty of murder and the judge sentenced him to 25 years to life in jail.

Of course, the diatom test for drowning is not a perfect test. According to the 'criterion of concordance', it's not just the presence of diatoms in the bone marrow that conclusively indicates drowning. The relative abundance of diatoms in the river, stream, lake, or pond water where the presumed drowning took place needs to be similar, taking into account that the largest diatoms may not have been able to pass through the alveolar tissue into the bloodstream. This taxonomic matching is necessary to dispel the view of some pathologists that diatoms accumulate in body tissues over the course of our life anyway.

Besides diatoms, other microscopic, unicellular algae are also of potential forensic value. Dinoflagellates responsible for many toxic 'red tides' and the chrysophyta or golden algae (e.g. *Mallomonas caudata* in the Connecticut pond example, above) are especially notable. Macroscopic filamentous and thalloid algae, including many green, brown, and red algae, are also of such value. And filamentous cyanobacteria are responsible for the release of dermatoxins, neurotoxins, or hepatotoxins following blooms in water bodies (see Chapter 6).

Yet, despite the long human history of dumping bodies in the ocean, macroscopic algae have rarely featured in forensic cases. Nevertheless, algae, including macroscopic seaweeds, to some extent colonized 36% of human bone cases recovered from the shoreline of ocean waters near Massachusetts. These encrustations offer potential as forensic evidence in post-mortem investigations.[15] In a 2019 drug bust case, a large pile of seaweed encrusting bags of drugs was the giveaway for investigators searching for stashed cocaine, ecstasy, and other street drugs after a drug smuggler's yacht hit a reef and ran aground in the Abrolhos Islands of Western Australia.[16]

Filamentous algae on his T-shirt were the undoing of John C. Hoeplinger, a 35-year-old business executive.[17] At 4:55 a.m. early in the morning of 7 May 1982, he called in a report to the police, saying that he had found his wife Eileen's dead body outside their house in affluent Easton, Connecticut. Hoeplinger claimed that after finding his wife missing in the middle of the night, he had searched for her and found her body on the side of the driveway. He claimed that he brought her inside, placing her body on the sofa in the family room. He said that he tried to clean up the blood so that the couple's two young children would not become upset when they got up. However, several clues, including blood-spatter evidence, algae, and the appearance of some of the plants growing alongside his driveway, cast doubt on his story.

When the police arrived, they found Eileen's body on the sofa in the family room. There was an extensive amount of blood on the sofa, her clothing was in disarray, there were abrasions on her chin and back, and there were gravel and plant fragments on her blouse and skin. Blood splatter patterns on the sofa and adjacent furniture suggested she had died on the sofa after being brutally beaten. Police brought in forensic investigator Dr Henry Lee to work the case. He found blood tracks in the driveway and into woods and blood on the upper surface of the leaves of the herbaceous plants in the woods at the edge of the driveway, and that the plant stems were bent to face away from the drive. All of this blood-splatter and plant evidence suggested to Lee that the killer had carried the bleeding body out from the house, and down the drive before dragging it into the woods. The killer then changed his mind about dumping the body in the woods. The blood trail shows he carried the body back inside and put her back down onto the sofa before calling the police.[18]

The police still needed the murder weapon to better show that their suspect, John Hoeplinger, had committed the crime. Another trail of blood droplets led outside to the deck, onto the deck steps, across the lawn, and down to a pond. There was blood on rocks at the edge of the pond. In shallow water at the edge of the pond, investigators found a brick that had the victim's blood and hair on it. The brick was the same colour and size as bricks lining the driveway in the front of the house. Lee deduced that the brick was the murder weapon used to beat Hoepinger's wife to death on her sofa. But Hoeplinger denied it all, continuing to suggest that the unknown intruder used the brick. He claimed this intruder must have carried out the murder.

In a key development, investigators found a wet T-shirt hanging on the deck railing. This T-shirt was the same size and design as others of Hoeplinger's found in his clothes closet. Eileen's blood was on the T-shirt, but Hoeplinger claimed that the blood got onto it after he found his wife's body. It got wet, he said, because he rinsed it out. However, using scanning electron microscopy, the technicians at the Connecticut Forensic Science Laboratory identified green stains on the T-shirt as filamentous green algae that came from the pond. The conclusion was that he washed his shirt off in the pond after he dumped the brick, not realizing that, in the spirit of Locard (Chapter 2), he was picking up the tell-tale algal filaments at the same time. The filamentous algae on the T-shirt helped link the killer to the murder weapon and the place where he dumped it. Jurors found Hoeplinger guilty of first-degree murder. He later died in prison.

* * * * *

As I mentioned briefly in the discussion of body dumps, fungi and their spores can also provide important forensic evidence.

In what has been called the Australian Lindbergh case, Stephen Bradley kidnapped and murdered eight-year-old Graeme Thorne in 1960 in Sydney. Fungi and plant fragments provided important forensic evidence in this case.

Bradley kidnapped the boy at about 8:30 a.m. on 7 July 1960, while he was on his way to school in Bondi, a suburb of Sydney. He kidnapped the boy because he had read in the local newspaper that Graeme's parents Bazil and Freda Thorne had recently won £100,000 of lottery money. Bradley had demanded a ransom of £25,000, but perhaps realizing that the police were closing in, he killed the boy, dumped the body, and escaped from the country.

On 16 August, a month after the boy's kidnapping, investigators found his body wrapped in a rug on a vacant allotment. The body was in the advanced stages of decomposition but showed evidence of a head injury, skull fracture, and asphyxia that together had caused his death. Plant pathologist Neville Wright from the School of Agriculture, University of Sydney found that Graeme Thorne's leather shoes had abundant fungi on the heels. Wright identified four kinds of fungi and one in particular, a yellow-green mould (*Aspergillus repens*), had reached the growth stage at which fruiting bodies (perithecia) had developed, allowing the release of spores.

Perithecia are cylindrical, round, or flask-shaped fruiting bodies most commonly found on the group of mould fungi known as Ascomycetes or sac fungi. The fungus releases its dust-like spores from a pore in the perithecium. Given the conditions of the body's location, Wright estimated that it would have taken a minimum of three weeks for the fungi to develop to this stage. The perithecia had started to break up, releasing their spores, which occurs a further three weeks after they appear. These fungi would not grow on shoes that were being used or worn, and they require a covered, warm, humid environment. Their presence and growth stage indicated a PMI of around six weeks, placing the victim's death within 24 hours of his kidnapping. Wright corroborated this estimate with additional evidence from nematodes he found on the boy's socks and blow fly larvae on his corpse.

Lead investigator Detective Sergeant Alan Clark also found plant leaves and seeds on the boy's clothing. Joyce Vickery, together

with other botanists at the Royal Botanic Gardens, Sydney identified these plant fragments as two garden shrubs: *Chamaecyparis
pisifera* var. *squarrosa* (Sawara cypress) and *Cupressus glabra* (Arizona
smooth-bark cypress). Neither of these plants are native to Australia, but many homeowners plant them as garden ornamentals.
While Sawara cypress was reasonably common in local gardens,
Arizona smooth-bark cypress was rare. These two plants only
occurred growing together in the front garden of the suspect's
house in the north-eastern suburbs of Sydney, about 3 km from
where the body was found. Additional trace evidence of human
and dog hair and some pink mortar fragments also linked the
body to the suspect's house and car. It appears that the boy's body
had been lying under the house before being moved to the dump
site.

By mid-October, the police had enough evidence to arrest
Stephen Bradley. He was on board the SS *Himalaya*, a ship bound
for England, but was taken into custody when the ship docked en
route in Colombo, Ceylon (now Sri Lanka), and was returned to
Australia for trial. The court found Bradley guilty on 29 March
1961, and the judge sentenced him to life imprisonment for the
murder of Graeme Thorne. He died of a heart attack in jail in 1968,
aged only 42.

The fungal and plant evidence in this case was important in linking the suspect, the suspect's house, and the body of the victim.
The case received a tremendous amount of publicity for several
reasons. It was the first kidnapping and ransom case in Australian
legal history. At that time, kidnapping did not even make up an
offence on the statute books. (The Australian Parliament subsequently rectified this legal loophole by passing an amendment to
the Crimes Act in December 1961.) With the mix of compelling

evidence, this case also ushered in a new era of forensic science to the Australian legal system.[19]

Fungi, including those that provided key evidence in the previous case, are not plants, and evolutionarily they are more closely related to animals. Nevertheless, botanists have traditionally included their study within their remit. After all, a mushroom looks superficially more like a plant than an animal, and fungi produce spores that look similar to some pollen grains.

Fungi are a very diverse group of organisms whose cells have a membrane-bound nucleus, just like plants and animals (that is, they are eukaryotes). Fungi have been around for more than 600 million years, and there are thought to be more than 1.5 million living species, only about 100,000 of which have been scientifically described. The fungi are in their own kingdom in the tree of life, and include rusts, smuts, yeasts, mushrooms, sac fungi, penicillium, moulds, and a whole host of microscopic soil and pathogenic fungi. Fungi cannot photosynthesize, and so they need to scavenge their nutrients and glucose from other organisms. When associated symbiotically with certain algae, fungi form lichens, which can have some forensic value.

Forensically, fungi can play a role in estimating PMI (as in the Thorne case above), in indicating the time of deposition of a dumped body, in locating corpses, in poisoning cases, in assessing public health through mould growth, and in collecting trace evidence. Any time that an investigator observes fungi at a crime scene, they should consider them as potential evidence.

Edmond Locard provides perhaps the earliest example of the use of fungal trace evidence in a homicide case.[20] He wrote that on 8 June 1924, gendarmes found the corpse of a Mr Boulay following his disappearance nine days earlier. Examination of the corpse revealed that the back and sleeves of the victim's shirt and his hair had traces of anthracite coal dust along with mushrooms and yeasts typical of cellars. Mr Boulay was known to place bets with his concierge, Mr Tessier, who received customers in his cellar.

Examination of Mr Tessier's cellar suggested that he had recently cleaned it, but it retained traces of blood, coal dust, and the same mushrooms and yeasts found on the victim. The investigators also found a return metro ticket, dated the day of Mr Boulay's disappearance, to George V station, where Mr Boulay regularly travelled. All of this evidence linked the body to Mr Tessier's cellar, establishing it as the crime scene and Mr Tessier as the murderer. He was convicted of murder, sentenced to 10 years in jail, and made to pay 10,000 francs to the widow of the victim.

Fungal spores can have forensic evidentiary value, much like plant pollen. Evidence from fungal spores can help provide a 'picture of place', as they did in a horrific modern-day rape and murder case in Dundee, Scotland in 2010.[21, 22] Thirty-four-year-old Brenda MacEwen (not her real name), married mother of three children aged 11, 15, and 18, went missing on 25 February 2010. She had been at the Fat Sams night club that evening with her sister when she met serial sex offender 41-year-old Phillip Roebuck (not his real name). CCTV records showed her dancing with the suspect and then leaving the club with him. She was never seen alive again.

Two weeks later, police found MacEwen's body 50 metres away from the nightclub near the Ladywell Roundabout on North Marketgait road. Her body was at the base of a wall and covered with leaves and stems of ground ivy (*Hedera helix*) that the killer had pulled off the wall. Roebuck had beaten, raped, and strangled MacEwen before dumping her body. The cold February temperatures had preserved her body well, and extensive fungal growth on the hands and face was consistent with a two-week PMI.

The crime scene was essentially an ornamental border that had an exotic tree, various shrubs, and macerated woody mulch on the ground made from garden and park waste, which contributed to a unique plant mix.

Investigators found that the pollen and fungal spore taxa matched between samples collected from the suspect's shoes, the roundabout vegetation and soil, and the victim's clothing. But the taxa were different to those at nearby locations. For example, there were spores of 16 fungal taxa common to the suspect's shoes, the victim's clothing, and the crime scene. Of the 25 fungal taxa from the crime scene, 17 were also on the suspect's shoes. Three fungal taxa on his shoes had fruiting structures (i.e. spore-producing structures such as mushrooms) that would have required direct physical contact to pick up. The fungal spores were from taxa expected from the identity of the woody vegetation growing at the crime scene. Plants identified from the pollen were unique to the unusual, unnatural (in an ecological sense) assemblage of plants growing on the roundabout. This uniqueness was because of contributions from the mulch which retained pollen from plants that were in flower during the previous growing season.

In this case, the plant and fungal evidence clearly linked the suspect to the crime scene and the victim. The evidence against the suspect was overwhelming. In addition, the suspect already had 13 prior sexual offences from time spent in Ireland, and it was unclear how he had even moved to Scotland without appearing on the radar of the local authorities. On 3 June 2011, following a jury trial in the High Court of Edinburgh, Judge Lord Tyre sentenced

Roebuck to mandatory life in prison to serve a minimum of 20 years. During sentencing, the judge said to Roebuck, 'You have shown no remorse for what you did. The horror and terror that Brenda MacEwen must have experienced during your murderous attack on her can only be imagined.'[23]

Can fungi provide additional forensic evidence? Indeed, you would be forgiven for thinking you've come across a buried body if you saw the grey, finger-like stroma of a dead-man's fingers fungus (*Xylaria polymorpha*) poking up through woodland soil. You haven't found a body. As a saprophytic fungus, it is growing on decomposing wood, not on a cadaver. However, fungi can help in the search for clandestine graves. There's a group of so-called cadaver fungi or corpse-finder fungi that have this potential use.

These types of fungi include mushrooms of disturbed ground that take one to two years to produce fruiting bodies, for example, the shaggy ink cap (*Coprinus comatus*) and some morels (*Morchella* spp.). There are also 'ammonia fungi' that will fruit following stimulation from the release of nitrogen-rich decomposition products. These include the agaric fungus *Hebeloma vinosophyllum*, which was found in a Japanese study to be growing over buried corpses of a dog and a cat. Similarly, there are some reports of the rooting poison pie agaric *H. radicosum* and the so-called corpse-finder *H. syrjense* associated with corpses. But few if any of these reports are properly documented, and so the value of corpse-finder fungi remains anecdotal.[24]

The use of fungi as tools for PMI estimates remains poorly developed[25] although the Graeme Thorne case illustrates its potential. Unfortunately, the so-called map lichens, such as *Rhizocarpon geographicum*, which have been used to date old gravestones based upon the size of their colonies, don't grow on calcareous materials such as bones. However, several fungi, including moulds and some lichens comprising the necrobiome (the organisms associated with decaying corpses), can colonize and grow directly on the skin of decomposing bodies (e.g. species of *Aspergillus, Candida, Mucor, Penicillium*), or skeletonized bones (e.g. the orange lichens *Caloplaca* and rim lichens *Lecanora*). These fungi grow under specific environmental conditions, and their growth rate has potential for PMI estimates. One estimate suggests that it takes three to seven days before fungi can colonize a corpse to allow a PMI approximation after fungi are first observed.[26]

In 2006, police recovered the bodies of two sex workers from water in Suffolk, England. Fungal mycelium—the vegetative hyphae of fungi—of *Fusarium, Geotrichum,* and *Mucor* species, and of *Pythium* (an oomycete, technically not a fungus), covered their bodies—one more so than the other. Mycologists did not know the growth rates of these fungi when growing under water. Nevertheless, forensic investigators estimated two and five weeks' submersion, respectively, for the two women's bodies. These estimates agreed with information on how long they had each been missing and helped investigators trying to track down the serial killer responsible for the deaths.[24]

* * * * * * *

We have all seen the mould that grows on spoiled food. In 2013, a child died from neglect after being left with two siblings by their mother in a locked flat in London. The mother said that she had only left them for the weekend. The size of colonies of *Aspergillus niger* and *Geotrichum candidum* moulds on cooked food in the flat suggested 10 to 14 days. Investigators came up with this estimate by measuring the size of the fungal colonies and comparing them with growth rates estimated from published lab experiments on artificial media under similar temperatures. Confronted with this evidence, the mother pleaded guilty to child neglect.[21]

The forensic identification of mould can play an important role in civic cases related to human health. In a 1997 civic case, the judge assessed $11 million in damages against a negligent construction company for mould that was causing health problems in the county courthouse of Martin County, Florida. These 'toxic moulds' or mildews can produce mycotoxins associated with respiratory health issues, colloquially known as 'SBS' or sick building syndrome. Specific health issues include infections, allergic bronchopulmonary aspergillosis, allergic fungal rhinosinusitis, hypersensitivity pneumonitis, and asthma, all of which have a well-accepted medical link. The fungi responsible for these maladies are in three major fungal groups, namely the Ascomycota, Deuteromycota, and Zygomycota. The microscopic spores of these fungi are ubiquitous in household dust, particularly under humid conditions. Investigators can get minimum-age estimates of these moulds from colony sizes. These estimates can be of value in landlord–tenant disputes and in assessing the validity of insurance claims for water damage.

The monetary size of these toxic mould claims can be staggering—$4 million for a Texas homeowner for a property damage claim and $18.5 million in a California mould contamination case. The entertainer Ed McMahon accepted a $7.2 million settlement in his mould claim against his insurance company and environmental clean-up contractors for personal injuries, property damage, and respiratory illness suffered by his dog.[27] While some respiratory health issues related to mould exposure have a solid medical basis, others, especially the airborne mycotoxins supposedly produced by the greenish-black mould *Stachybotrys chartarum*, are considered by some to be little more than junk science.[28]

As we have seen from the cases in this chapter, microscopic plant (pollen), algal (diatoms), and fungal (hyphae, fruiting bodies, or spores) structures can all contribute valuable forensic evidence to investigators. Such evidence would not be readily apparent at a crime scene, except perhaps if crime scene investigators observed obvious macrofungal fruiting structures like mushrooms. Rather, investigators must take the initiative to sample 'les poussieres', secure in the knowledge that microscopic examination of spores, pollen grains, and unicellular algae may reveal valuable evidence. This microscopic material is evidence that a perpetrator cannot know that he or she is transferring and spreading around. In the next chapter, our investigations will scale down even more, to the molecular level, as we examine the importance of plant DNA evidence.

5

IT'S IN THE GENES

They asked me whether it was possible to link the pods to a particular tree the same way human fingerprinting could link blood or semen samples to an individual.

Geneticist Tim Helentjaris (1993) commenting on his analysis of plant DNA in the Maricopa murder case[1, 2]

You can bet that Mark Bogan wishes he had kept the bed of his pickup truck clean. How was he to know that DNA from a few seed pods found in his truck would provide the key evidence to convict him of murder? Forensic scientists used DNA extracted from plant material in 1992, for the first time, to place a suspect at the scene of a crime in the highly publicized 'Maricopa' case. The method used was crude by today's standards, but good enough. Investigators showed that the seed pods found in Bogan's truck came from a tree next to where police discovered the body of his victim. The DNA of these seed pods matched a particular tree at the crime scene and didn't match other trees of the same species elsewhere.

Counsel introduces molecular evidence from human DNA into evidence fairly routinely these days. But the use of plant DNA evidence has lagged behind. Why?

Scientists developed and began using human DNA profiling in forensics before plant DNA technology was developed.[3] In 1984, British scientist Sir Alec Jeffreys at the University of Leicester realized that the great variability in human DNA meant that each individual had a unique genome. This uniqueness was considered analogous to the well-known fingerprint evidence that police have routinely used in criminal investigations since the time of Edmond Locard, hence the early use of the term DNA fingerprinting. The issue facing Jeffreys was how to sample the small amounts of DNA contained in cells and how to conduct the analysis.

In fact, less than 0.1% of the chromosomes in human DNA are unique to an individual. For that matter, researchers have known since 2005 that we share 99% of our DNA with chimpanzees, our closest living relative. But since human DNA is large, 0.1% of it is enough to create a unique profile for each of us.

Strings of hundreds of copies of four bases—adenine (A), thymine (T), cytosine (C), and guanine (G)—make up each of the two complementary strands of all DNA. These are not the familiar chemical bases like baking soda, ammonia, or milk of magnesia but nucleobases—nitrogen-containing molecules that are the fundamental building blocks of the genetic code. The four bases are joined to each other in a distinct pattern: adenine bases on one strand pair with thymine bases on the other, and cytosine bases pair with guanine bases. Human DNA found in the nuclei of almost all our cells has 3 billion such base pairs. Each base joins with a 5-carbon sugar molecule and a phosphate group to form a nucleotide. A gene is a sequence of nucleotides along DNA strands that the cell interprets to synthesize proteins. But there are lots of non-coding sections of DNA too.

The 0.1% of human DNA that confers our unique identity includes short sequences of bases that are repeated over and over again. These sequences are known as short tandem repeats (STRs). For example, in humans some of the most common STRs are based upon adenine repeats. STRs are passed from one generation to the next, often with a gain or loss of their repeats. A parent may have 10 copies of a particular STR, while their child may have even more. Nonetheless, these repeats don't affect how a gene works.

Jeffreys determined that he could use the number of repeated bases at STRs across the genome to generate a unique 'DNA fingerprint'. The problems of this method at the time were limitations associated with analysing small amounts of DNA and handling often contaminated or degraded DNA samples (this is still a problem). The first of these problems was solved by the chemist Kary B. Mullis when he developed the polymerase chain reaction (PCR) in 1985. This technique allows the 'bulking up' of small amounts of DNA to provide a large enough sample for analysis. The same base structure and rules of base-pairing is found in the DNA of all animals and plants, but forensic applications of DNA profiling were initially more of a priority for the development of human rather than plant or (non-human) animal methodology.

The Colin Pitchfork murder trial in the UK in 1987 was the first human DNA forensic investigation and attracted a lot of attention.[4] Following his discovery of DNA pattern recognition based upon STRs, Jeffreys had been investigating paternity and immigration cases. Then, in 1986 the police asked him to investigate the rape and murder of 15-year-old Dawn Ashworth in Narborough, Leicestershire, England and a similar earlier cold case from 1983 of the rape and murder of 15-year-old Lynda Mann. In both cases, the young girls went missing while walking home. Police found their bodies a day or two later.

Could DNA samples from semen collected from their bodies provide a match with a suspect? Surprisingly to the police, Jeffreys found that the DNA samples didn't match the DNA of the primary suspect, Richard Buckland—so the police excluded him from the investigation. Subsequent analysis of the DNA from more than 5,000 blood and saliva samples from 17- to 34-year-old men in the Leicestershire area also did not find a match. The case was stonewalled.

However, the police caught a break. While drinking in a bar, a source overheard a man named Ian Kelly bragging that he'd been paid £200 to pose as his friend, local baker Colin Pitchfork, to provide a false DNA sample. When police subsequently arrested Pitchfork on 19 September 1987, and tested his DNA, it matched the DNA obtained from both crime scenes. Presented with this evidence, he confessed to both murders. The court found him guilty and the judge sentenced him to a life imprisonment term of 30 years, which was later reduced on appeal to 28 years.

Forensic scientists now commonly use DNA profiling, as it is known today, to determine the probability of a match between DNA samples. Authorities have developed several human DNA databases in different parts of the world. Forensic scientists can tap into these databases to seek a potential DNA match from an unknown sample. Specialized software uses mathematical algorithms to conduct 'probabilistic genotyping' to determine 'match statistics' that give probability values for how much better a known or reference profile (e.g. from individuals already in the database) explains the evidence sample than a profile chosen at random. These are the statistics that forensic scientists then present to the attorneys or in court at trial.

The US FBI created a human Combined DNA Index System (CODIS) in 1990. This system uses 20 different STRs across the human genome in its National DNA Index System (NDIS). More than 14 million offender profiles, more than 3.7 million arrestee profiles, and almost one million forensic profiles were stored in CODIS as of November 2019.[5] Similarly, the National Criminal Investigation DNA Database (NCIDD) in Australia, which is based upon 18 STR loci and a sex gene, has more than 837,000 DNA profiles as of July 2018. Many other countries, including Brazil, Canada, China, France, Germany, India, Israel, Kuwait, New Zealand, the Netherlands, Sweden, and the United Kingdom, have similar national DNA databases. These DNA databases must reflect the local population if the statistical evaluation of the evidence is to be valid. Interpol's DNA database, created in 2002, contains more than 180,000 profiles from 84 member countries. Using these databases involves the application of statistical probabilities, and it is easier to exclude a suspect than match suspects. Early on, admissibility of DNA profiling evidence in trials involved overcoming *Frye* and *Daubert* challenges by opposing counsel, as well as the Bible challenge we'll look at shortly. These challenges for forensic scientists remain as scientists bring in improved and new technology.

Forensic scientists did not use molecular evidence from plants until 1992, and even today, its use is uncommon for several logistical reasons. For a start, plant DNA doesn't have the background of databases to draw upon that human DNA studies have. Nevertheless, unlike human DNA profiling, which specifically seeks to tell apart individuals of a single species, *Homo sapiens*, there are a greater variety of ways in which DNA can be useful for botanical forensics.

Just as in human DNA profiling, investigators may need to discriminate among individuals, cultivars, or varieties of a single plant species. For example, investigators can help track the source of illicit drugs such as cannabis. Or, in civil cases, investigators may seek to determine if a cultivar or variety of a crop is the same as or different to another one. The latter is important, for example, when a biotech company suspects that a farmer is illegally growing plants of a patented GM crop variety. Seed companies that genetically engineered these crops get very upset about farmers growing them without a licence. They wish to protect their legal rights on the patented crop, recoup their financial R&D investment through sales, and make a profit.

These cases of DNA profiling are thus similar to human DNA profiling. In fact, the first plant DNA case, the Maricopa case, was analogous to human DNA profiling precisely because the prosecution sought to determine whether seeds in the suspect's truck were from a particular tree growing where the body was found or from other trees of the same species commonly found elsewhere in the region.

Forensic scientists can also use DNA to identify plant fragments found associated with a crime scene. In these cases, the forensic scientist wants to identify an unknown plant species. Identifying plant fragments carried from a crime scene on a perpetrator's clothing, shoes, or vehicle can help place them at the scene if the plants match. This application of identifying different species from their DNA is known as DNA barcoding.

Finally, forensic scientists can detect adulterations of food, drink, and medicinal products using DNA markers for known clones, cultivars, or domesticated varieties (landraces). For instance, scientists can test the purity of an expensive olive oil to confirm its provenance or help identify a fraudster.

* * * * *

But let's come back to that first use of plant molecular data, in the so-called Maricopa case that led to the conviction of Mark Bogan.[6, 7] Desert trees formed critical evidence in this case. On the morning of 3 May 1992, Tim Faulkner, a dirt bike rider, found a woman's naked body lying face down in a remote area of brush outside the Caterpillar Proving Grounds, an abandoned factory in Maricopa County, Arizona. As the woman appeared to be dead, Faulkner immediately rode home and phoned the police. After arriving on the scene, investigators quickly determined that someone had dumped her body after they had strangled her to death. Wet blood on her body suggested that her death was recent. The police identified her as Denise Johnson, a sex worker from nearby Phoenix.

Matted vegetation of the grass near Johnson's body (a straight-forward botanical clue) indicated that someone had dragged her body a short distance to where Faulkner found her. Her killer had bound her body with cloth around her neck and left wrist, a shoelace around her left ankle, and braided wire around her right wrist and right ankle. Her clothing was scattered around. A pager found at the scene was registered to Earl Bogan, but it was used primarily by his son Mark. Police arrested Mark Bogan as a suspect.

While the police were investigating the scene, Chad Gilliam, a local man, came forward and stated that he had passed by the Caterpillar Proving Grounds while driving home from a party at 1:30 a.m. that morning. At that time, he noticed a white pickup leaving the scene, running a stop sign, and travelling 'pretty quick'. It turned out that Mark Bogan drove a white pickup matching Gilliam's description.

So far so good, but the investigators needed more evidence to place Bogan at the crime scene. Investigators then found two seed pods of a blue palo-verde tree (*Cercidium floridum*, also called *Parkinsonia floridum* by some taxonomists) in the bed of Bogan's pickup truck. This tree is native to the Sonoran Desert region of the south-western United States. The common name, blue palo-verde, means 'green pole or stick' in Spanish. Individual blue palo-verde trees commonly grow 15–20 feet in height, with blue-green photosynthetic bark and pinnately compound leaves, spines on the branches, and small, fragrant, bright-yellow pea-like flowers appearing in March to May. The 1.5–4-inch-long, tan-coloured, bean-like pods contain one to eight flat seeds that dry on the plant before being shed in the autumn. A branch of a blue palo-verde tree dipping low over the driveway near Johnson's body had a fresh abrasion, perhaps from a vehicle. Crime scene investigators designated this tree as 'PV-30' and it played a prominent part in the investigation—making this tree perhaps the best-known blue palo-verde tree in the world.

Bogan lived an 18-minute drive from the crime scene with his wife, Rebecca Franklin. She testified that he had been drinking heavily the evening of 2 May 1992, before going out. He woke her up when he returned home that night at 2:03 a.m., and she noticed recent scratch marks on his face he claimed were from a bar fight. She also testified that she saw braided wire in his truck prior to that evening that matched wire found on Johnson's body. However, investigators did not find any wire in his truck when they searched it on 5 May 1992.

An irony of her testimony was that Bogan's lawyers appealed to exclude Rebecca Franklin's testimony based on anti-marital fact privilege (i.e. spouses can't testify against each other). But the state's marriage records showed that Bogan was actually still married to a Teresa Bogan—indeed, she testified that they had been married since 1982. Bogan's supposed marriage to Rebecca Franklin was invalid and bigamous, allowing her important testimony for the state to stand.

When questioned by detectives, Bogan admitted picking up a hitchhiker matching Johnson's description. He claimed that they had sex in his pickup, but when they argued, he made her get out of the truck. He added that she 'swiped' his wallet, pager, and items from the dashboard of the truck before attempting to run away. He chased her down, and she scratched his face while he fought to retrieve his property. He said that he didn't realize until the next day that he hadn't retrieved the pager. Despite the arguments and the fight, he denied being at the crime scene, and he denied killing her.

The police had some sound evidence, but lead detective Charlie Norton needed more to convincingly place Bogan at the scene where they had found Johnson's body. He must have wondered if DNA profiling evidence already established as admissible in courts as human evidence could be used with the blue palo-verde seed pods found in the suspect's truck. He enlisted help from Dr Timothy Helentjaris, a professor of molecular genetics at the nearby University of Arizona. In a later interview, Helentjaris said, 'They asked me whether it was possible to link the pods to a particular tree the same way human fingerprinting could link blood or semen samples to an individual.'[1] He did exactly that.

Helentjaris used a DNA method known as randomly amplified polymorphic DNA (RAPD) analysis. He first blind-tested 12 trees from around the site—meaning that police provided him the samples without any identifying labels as to which tree their CSIs collected them from. He determined that they all had unique genetic profiles. He then did a RAPD analysis of the two seed pods from Bogan's truck and the damaged tree PV-30 near where Tim Faulkner found the body. The seed pods from Bogan's truck had the same profile as the damaged tree. Then he carried out blind tests on 18 trees of the same species growing outside the crime scene, including Bogan's old haunts. Again, each had a different RAPD molecular profile and didn't match profiles of trees at the crime scene or the seed pods found in the truck.

But would the court in the Maricopa case accept DNA evidence provided by Helentjaris? I described the standards required for admissibility of scientific evidence in Chapter 2. The judge required a *Frye* hearing because of the new DNA technology that the prosecution wanted to introduce into evidence in the Bogan trial. Three experts testified, agreeing that the scientific principles underlying RAPD-DNA analyses were valid and met the general acceptance standard. The method was 'generally accepted' by the scientific community. Even the defendant's expert, renowned geneticist Dr Paul Keim of Northern Arizona University, acknowledged the soundness of the technology represented by RAPD analysis and that Helentjaris had conducted the analysis appropriately. The *Frye* hearing concluded that the results of Helentjaris's RAPD analysis were admissible as evidence.

The second criterion after clearing the *Frye* standard is that presentation of the evidence must be subject to a 'foundational showing' following standards established in an earlier case involving human DNA argued in the Arizona Supreme Court.[8] This case established the need for an accepted standard in the scientific community to declare a match between samples following DNA testing. Here, Helentjaris and Keim disagreed on the odds of a random match in this case. Helentjaris testified that the odds were one in a million, whereas Keim estimated the odds at one in 136,000. The population of Maricopa County, Arizona was 2,122,210 in 1990, so Keim's odds would have included about 13% of the men in the county, all other things being equal.

The trial judge followed an earlier precedent-setting ruling in *State vs. Bible*[8] and decided that statistical evidence on the probability of a random match was inadmissible. However, Helentjaris was instead allowed to testify that the seed pods recovered from Bogan's truck and those from blue palo-verde tree PV-30 'were identical', that he was 'quite confident in concluding that these two samples . . . most likely did come from [PV-30]', and that the samples 'matched completely' and 'didn't match any of the [other] trees'. Helentjaris was confident in his analysis, testifying that the DNA from PV-30 would be unique among 'any tree that might be furnished' to him.[6]

The judge didn't allow Helentjaris to put a statistical probability onto the results, as they do in human DNA evidence derived from semen samples, because of the limited number of trees sampled. By comparison, forensic scientists use human DNA evidence based on comparisons with databases of hundreds of thousands of people. At trial, Helentjaris said, 'in my professional opinion it was highly likely the seed pods came from that tree and it was highly unlikely that they came from any other tree by chance'.[1]

The plant DNA evidence was compelling enough to lead to a murder conviction in the Maricopa case. Using RAPDs was rudimentary compared with today's standards and it is not a method of DNA profiling. RAPDs are nevertheless accurate at telling apart organisms whose genomes have not been mapped or fully sequenced—as is the case for the vast majority of plants.

Researchers developed RAPDs in the late 1980s, and they appeared in scientific journals in 1990. At the time of the Bogan trial, Helentjaris had never used this method before for forensic purposes. It works by identifying and detecting distinctions in DNA samples as part of the polymerase chain reaction (PCR) technology that researchers use to generate exponentially large numbers of identical DNA copies from a tiny sample. RAPD technology uses laboratory-synthesized chemical primers of arbitrary sequences to enable DNA replication. A primer consists of a randomly selected short, single-strand sequence of 8–10 nucleotides that binds with its equivalent set of nucleotides (its homologue) on the DNA strand, amplifying that particular section of the DNA.

The equivalency of nucleotide sequences between the DNA strand and the primer means that during PCR, multiple copies of just the section of DNA catalysed by the primer are produced. A number of chemically different primers will therefore generate multiple copies of the particular sections of DNA they each represent. Here, it doesn't matter that the investigator doesn't know what genes these sections of DNA code for, if they are genes at all. Nevertheless, the investigator can compare their sequence variability among samples from different individuals.

Several years ago, Danny Gustafson, one of my research students at the time, used RAPD analysis to distinguish among samples of two prairie grasses: big bluestem (*Andropogon gerardii*) and Indian grass (*Sorghastrum nutans*). He sought to determine the suitability of different seed sources to use in prairie restorations.[9] Similarly, in forensic applications, if DNA from two individuals (e.g. from two different blue palo-verde trees in the Maricopa case) is different, then the DNA segments extracted using a particular primer can produce different RAPD profiles. A particular primer is only useful if the DNA segments that it produces vary among individuals—and they may not, so scientists usually test several primers for each application. In the Maricopa case, Helentjaris could distinguish among the 12 blue palo-verde samples with which he was initially provided using just two primers. As he expanded his sampling to 28 trees, he used seven primers to create 47 markers.

* * * * * * * *

As methods of DNA profiling have advanced, each new method has had to be accepted by the courts. RAPDs were the first method, followed by AFLPs, microsatellites (simple sequence repeats or SSRs, and short tandem repeats, STRs), and SNPs (single nucleotide polymorphisms, or 'snips').

AFLPs (amplified fragment length polymorphisms) are a more focused method than RAPDs and use specific primers to amplify a targeted subset of DNA fragments. Investigators have subsequently used AFLPs to distinguish among tomato seeds (*Lycopersicon esculentum*) from gastric contents (see Chapter 3) based upon how the tomatoes were prepared for cooking.[10]

Microsatellites can be used for identification by counting the number of repeated bases in a short stretch of DNA to generate a unique DNA fingerprint. In one case, investigators attempted to link a suspect to a burial site through DNA microsatellites extracted from live oak (*Quercus geminata*) leaves taken from the suspect's

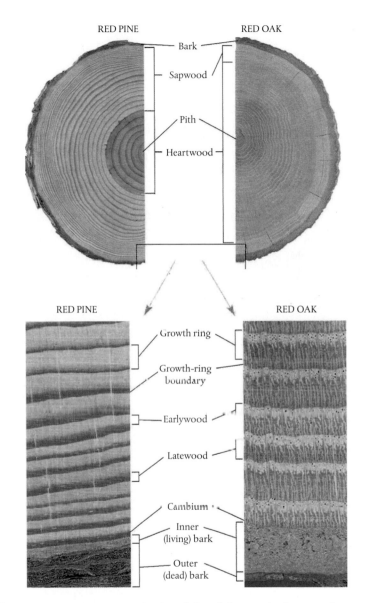

Plate 1 Stem cross-section (top) and detail (bottom) of a typical softwood, red pine (*Pinus resinosa*) and a typical hardwood, red oak (*Quercus rubra*).

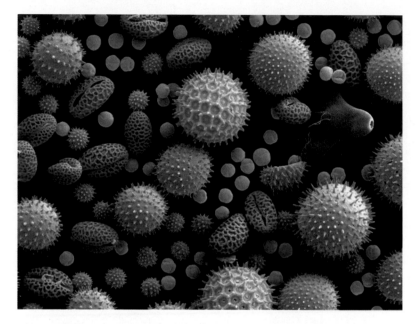

Plate 2 Pollen grains from a variety of plant species: sunflower (*Helianthus annuus*, small spiky sphericals, colourized pink), morning glory (*Ipomoea purpurea*, big sphericals with hexagonal cavities, colourized mint green), hollyhock (*Sidalcea malviflora*, big spiky sphericals, colourized yellow), lily (*Lilium auratum*, bean-shaped, colourized dark green), evening primrose (*Oenothera fruticosa*, tripod-shaped, colourized red), and castor bean (*Ricinus communis*, small smooth sphericals, colourized light green). Notice the varied ornamentation that helps with insect pollination. × 500 magnification.

Plate 3 Diatoms. Coloured engraving of an assortment of diatoms by the 19th-century biologist Ernst Haeckel.

Plate 4 *Nepenthes rajah*, the giant pitcher plant, has giant urn-shaped traps that can grow up to 41 cm high and 20 cm wide. This endangered plant is endemic to serpentine soils in Malaysian Borneo, where it is CITES Appendix I-listed.

Plate 5 The stunningly beautiful ghost orchid (*Dendrophylax lindenii*). This rare orchid is considered endangered in Florida and is CITES Appendix II-listed, but was being illegally poached from the Fakahatchee swamp, a nature reserve close to the Florida Everglades.

car and trees at the crime scene. In this case, the DNA-based identity of the trees did not provide a match or evidence against the suspect but proved the utility of the method.[11]

Scientists first showed the forensic value of microsatellites on botanical samples in the United Kingdom in 2001. Police found growth chambers that were being used for growing and cultivating marijuana in a warehouse in South Yorkshire. The warehouse suggested a large-scale marijuana operation, and the UK Forensic Science Service at the time described it as the 'most sophisticated ever seen in the UK'.[12]

Nearby, police found a large amount of marijuana in a private residence. Was the house used for sale and distribution of the marijuana from the warehouse? The police needed to show that marijuana in the house came from the warehouse, and so they enlisted help from Adrian Linacre at the University of Strathclyde, Glasgow. He tested nine plants from the warehouse, plus several plants seized from the house. Linacre extracted DNA and amplified it using polymerase chain reaction (PCR) amplification of an STR locus. All samples shared the same STR type that was itself very rare in a comparison, pre-existing marijuana database.

The court accepted this evidence and did not challenge it. This case set a legal precedent in UK courts for the admissibility of microsatellite DNA evidence, proving beyond a reasonable doubt that the marijuana was grown in the warehouse before being transferred to the house for distribution. The judge sentenced three male suspects to more than 28 years in prison for conspiracy to produce marijuana.

A murder case, in 2001, showed for the first time how the application of SSRs of mosses can provide key evidence in bringing the guilty parties to trial.[13]

Investigators found a missing man's body in a forest in Finland in September 2001. The police arrested three suspects, but they didn't have any blood or other evidence tying them to the crime scene. Witnesses last saw the victim alive three months earlier, leaving a café in a car with the suspects. All were known to the police as criminal associates—they knew each other and were partners in crime. Police found the victim's body 5 km from the café. Crime scene investigators identified plant material from samples found on the suspects' shoes, clothes, and in the car as three species of moss: *Brachythecium albicans*, *Calliergonella lindbergii*, and *Ceratodon purpureus*. Investigators noticed that colonies of these mosses also occurred at the crime scene near where the body was found. The police needed to know—were these plant samples pieces of the same plants found growing at the crime scene?

Helena Korpelainen and Viivi Virtanen, botanists from the University of Helsinki, took on the case. They used PCR analyses, using 10-base RAPD primers and 17- or 18-base SSR primers of two of the mosses, *B. albicans* and *C. lindbergii*. Korpelainen and Virtanen chose these two mosses for analysis as they both reproduce clonally, so fragments collected in evidence from suspects should match plants growing at the crime scene—especially when compared with samples from colonies of the same moss species collected elsewhere. The two botanists excluded the other moss, *Ceratodon purpureus*, from analysis as it more commonly reproduces sexually, so different plants from the same location are likely to be genetically different. Molecular analysis of a sexually reproducing moss such as *C. purpureus* can be done but would require a more sophisticated approach than time allowed (no surprise, the detectives were in a hurry). Results of the SSR DNA analyses showed a genetic match between samples collected from the suspects and the two tested mosses growing near the body at the crime scene. Prosecutors presented DNA evidence from the mosses in court, and the jury found all three suspects guilty at trial in December 2002.

SNPs are single nucleotide differences in stretches of DNA. For example, the nucleotide base cytosine may be replaced with the base thymine in a particular portion of the DNA of an individual. These variations arise during cell division as the DNA is copied, and we may think of these as molecular typos. SNPs are very common, occurring about once in every 1,000 nucleotides, depending upon the species and genetic diversity, and can be unique to individuals. The Olive Genetic Diversity Database (OGDD), used to authenticate olive oil (*Olea europaea*) provenances, is based on genetic microsatellite marker profiles (see below). And scientists with the Phylos Galaxy project (also see below) are using SNPs to tell apart cannabis varieties. Likewise, in a case we'll come to later, investigators identified illegally logged bigleaf maple (*Acer macrophyllum*) wood using SNPs.

These molecular methods each have pros and cons related to the amount and integrity of the DNA sample required, and the ease, logistics, and accuracy of the analysis. Each new method has to be accepted by the courts.

* * * * *

Whatever the need, a problem that forensic botanists have is that for plant samples, there are very few large DNA databases that law enforcement can draw upon. At best, there are the marijuana (*Cannabis sativa*) DNA databases that scientists have been developing.[14,15] These developing marijuana DNA databases allow law enforcement to use DNA profiling to determine where an illegal shipment of marijuana was grown or the source of a medical marijuana variety.

Commercially, the Phylos Galaxy project[16] uses DNA sequence data that have SNPs in the genetic code. The SNPs are used to produce an evolutionary, interactive 3D constellation map of the relationships among more than 3,000 cannabis and hemp varieties from more than 80 countries. Breeders can submit samples for analysis and in return receive a report showing where their sample fits in relation to other varieties in terms of its genetic makeup. This DNA database is like the commercial human DNA databases, such as AncestryDNA™ and 23andMe™, which provide indications of genetic ancestry. As the Phylos Galaxy database is updated with new samples, the location of a *Cannabis* variety compared with other varieties shifts. Users of the online database can search for varieties or growers and find genetic reports, flavour profiles, and pedigrees.

The potential forensic applications of the Phylos Galaxy database are clear. Forensic investigators can also use this sort of database to characterize cultivars. A link can be provided between production and street-level narcotics distribution.

Marijuana is just one plant species among 400,000 known worldwide that can be implicated or involved in criminal or civil cases. If the plant evidence is a piece of wood, or some resin or oil, how can a forensic scientist determine its provenance? For plants other than marijuana, the forensic scientist has to develop DNA databases almost on a case-by-case basis for particular species or groups of species.

For example, illegal logging and trafficking of endangered trees prompted the development of a DNA database to discriminate among natural populations of ramin (*Gonystylus bancanus*), a valuable but protected timber species from Malaysia and Indonesia.[17] Kevin Ng and colleagues at the Forest Research Institute in Malaysia used chloroplast DNA and STR markers to construct a database, allowing accurate discrimination among 17 natural populations of ramin. The hope is that this database will allow forensic scientists to determine the identity of processed wood samples, which would help law enforcement to combat illegal logging and trafficking. Similarly, scientists have developed limited DNA databases for other timber species, including European oaks, ashes, pines, and birches. These databases help with 'traceability'—the ability to trace the history, application, or location of an entity through recorded identification (see Chapter 7).

* * * * *

What can you do if you don't know the name of the plant that some botanical evidence is from? Distinguishing among different species or identifying an 'unknown' from DNA samples is more challenging than the single-species plant DNA applications I have described so far. To identify an unknown species, the investigator has to have a way to determine which of more than 400,000 plant species the extracted DNA is from. The approach taken is DNA barcoding, in which the DNA of a few commonly sequenced genes are extracted and sequenced and compared to a database of known samples.

Commonly sequenced and databased genes used for DNA barcoding in plants include the *rbcL* and *matK* plastid genes, *trnH-psbA* nuclear spacer genes, and the internal transcribed spacers (ITS) nuclear ribosomal spacer genes. Scientists chose these genes because they have highly conserved gene sequences, at least where the primers bind, meaning they change little within species in response to evolutionary selection pressures but are generally different between species. However, many plant species, such as many oaks (*Quercus* spp.), breed freely with each other. Extensive hybridization among plants over evolutionary time has complicated barcoding efforts, which is why barcodes are usually based upon more than one gene.

The plant working group of the Consortium of the Barcode of Life (CBOL) has recognized the hybridization issue and recommended the use of several genes, including the ones listed above. The US government-sponsored open-access GenBank database publicly archives nucleotide sequences of these genes.[18] Users can compare their unknown sequences with sequences in the database to search for likely matches using the BLAST (basic local alignment search tool) algorithm.

GenBank exchanges data daily with other nucleotide databases, including those of the European Molecular Biology Laboratory and the DNA Data Bank of Japan. Scientists can also potentially identify unknown DNA sequences using the Barcode of Life Data System maintained by the University of Guelph, Canada. This system included barcodes for 93,000 plant and fungal species by 2021, mined from databases worldwide, including GenBank.[19] The forensic potential of these databases and their utility to identify unknown plants from DNA extractions is tremendous and is expanding rapidly.

A forensic application of DNA barcoding is identifying fraudulent plant products. Tracking fraudsters in the food chain is particularly important to safeguard the integrity of economically and commercially important plant products, such as olive oils and wines. Accurate labelling of food products is now mandatory in most countries. In the first century, Roman philosopher and writer Pliny the Elder complained about the vagueness of the production and ingredients in liquor and wine: 'Alas, what wonderful ingenuity vice possesses! A method has actually been discovered for making even water intoxicated!'[20]

Similarly, have you ever wondered whether that expensive first-cold-pressed, extra-virgin olive oil on the supermarket shelf is really from an exclusive grower in Italy? What flowers make up that expensive boutique honey? What if the honey contained toxic residues? This contamination has actually occurred, as investigators have detected several toxic plant compounds in honey including hyoscyamine from *Atropa belladonna* and oleandrin from *Nerium oleander*.[21] Thankfully, DNA testing and technology can come to the rescue.

For example, the Olive Genetic Diversity Database has genetic microsatellite marker profiles along with morphological and chemical information of more than 200 olive oil (*Olea europaea*) cultivars from the major olive oil-producing countries, including France, Greece, Morocco, Italy, Portugal, Spain, Syria, Tunisia, and Turkey.[22,23] There is a web interface that allows users to readily retrieve and visualize biologically important information about cultivars on the database. Adulteration of olive oil, particularly extra-virgin olive oil, with inferior grade olive oil or even other plant oils is widespread. The Olive Genetic Diversity Database, along with chemical tests, can help determine the authenticity of olive oils and detect adulteration.[24]

Do you use herbal supplements? If so, *caveat emptor*. An example of the 'fraudster' problem was the explosive 2013 revelation by the *New York Times* that many herbal supplements 'are often not what they seem to be'.[25] An investigation led by Steven Newmaster at the University of Guelph Center for Biodiversity Genomics tested

44 herbal products sold by 12 companies supposedly containing 30 different plant species. The well-known echinacea (purple coneflower, *Echinacea purpurea*), ginkgo (maidenhair tree, *Ginkgo biloba*), and St John's wort (*Hypericum perforatum*) were among the herbal products tested. Newmaster and his colleagues used DNA barcoding to blind-test these herbal supplements in comparison with reference samples of 42 herbal species. To their astonishment, they found that 59% of the products were adulterated with plants not listed on the product label and 33% contained unlisted contaminants or fillers that could cause serious health problems to consumers.[26]

Product substitution occurred in 75% of the supplements. Common substitutions or fillers diluting the products included rice (*Oryza sativa*), soybean (*Glycine max*), and wheat (*Triticum* spp.). A more serious contaminant found in several herbal products was the agricultural weed Santa Maria feverfew (*Parthenium hysterophorus*), which contains the toxin parthenin. We know this toxin to cause dermatitis, nausea, flatulence, and respiratory malfunctions in humans. One sampled bottle of ginkgo contained only rice and another only walnut (*Juglans nigra*)—the latter a potential problem for people with nut allergies. A sample of black cohosh (*Actaea racemosa*), recommended for menopause symptoms including hot flushes, actually contained the related but toxic Asian baneberry (*Actaea asiatica*). Only two of 12 companies had 'clean' products as advertised.

The herbal supplement exposé highlighted by the *New York Times* is not the first time that scientists have used DNA testing to suggest truth-in-advertising problems in the supplement industry. The US Food and Drug Administration (FDA) had previously shown that many herbal teas contained herbs and ingredients that manufacturers had not listed on their labels. The most common non-labelled plants in a study of teas (*Camellia sinensis*) by Mark Y. Stoeckle and colleagues at the Rockefeller University, New York, included chamomile (*Matricaria recutita*), along with a fern (*Terpsichore* sp.), annual bluegrass (*Poa annua*), papaya (*Carica papaya*), and white goosefoot (*Chenopodium album*).[27]

Regulations in the supplement and herbal industry are not as strong as they could be, and strict enforcement is lacking because of limited regulatory agency resources. The *New York Times* quoted Susan Burgess, a spokesperson with the US FDA, saying that non-compliance with regulations by firms may be as high as 70%. The penalties for food fraud are so low that the supplement industry is basically self-policing, somewhat like putting the fox in charge of the henhouse.

* * * * *

DNA testing is particularly important in food security cases, such as the contamination of organic crops by GM crops. Threshold levels that trigger required labelling of GM content in crops and products are mandatory in many countries (China, the EU, Russia, Australia, New Zealand, Brazil, Israel, Saudi Arabia, Korea, Chile, Philippines, Indonesia, Thailand, and the UK) or voluntary in others (Canada, Hong Kong, the United States, South Africa, and Argentina). This labelling requirement means that there are legal reasons to determine accurately and precisely the occurrence of GM genes in a crop.

For example, molecular tests can determine expression rates of the *CP4 EPSPS* gene. This gene confers resistance to the common herbicide glyphosate (commonly marketed as Roundup™). Farmers use glyphosate to spray their fields, killing the weeds without killing the crop carrying the resistance gene. It did not surprise evolutionary biologists that it's more complex than the biotech companies thought (or publicly admitted). The worst weeds very quickly evolved resistance to glyphosate, allowing them to spread and compete in crop fields even more effectively than before. This rapid evolution has led to the development of GM crops resistant to multiple herbicide modes of action and the costly need for growers to use several herbicides together—a practice known as 'stacking'.

Regardless, if the herbicide resistance gene is carried in pollen from a GM crop into a non-GM organic crop of the same species, then the resulting seed may also carry the gene. Unlike commercial growers using GM crops, who have to buy new seeds from the agrochemical company each year, organic farmers thrive in part by collecting their own seeds to propagate their crop the next growing season. And if the seed is the crop, such as in soybean (*Glycine max*), wheat (*Triticum aestivum*), and corn (*Zea mays*), then GM-contaminated seed is not GM-free any more, leading to many truth-in-marketing problems, loss of organic certifications, rejection from markets banning GM products, and the chance of being overrun by the herbicide-resistant weeds that plague GM crop fields.

In the past, agrochemical companies have sued farmers for harvesting and sowing GM seeds. Biotech companies like Monsanto (a subsidiary of Bayer AG since June 2018) argue that patents protect their GM plants. They've been very successful in making this argument, having not lost a court case in more than two decades, suing more than 100 farmers for using GM seeds without a licence, and settling out of court in more than 700 cases.

A high-profile, precedent-setting example of the role of DNA testing in GM crop cases was the so-called David vs Goliath case between Canadian canola farmer Percy Schmeiser and biotech giant Monsanto. Canola (oilseed rape, one of several cultivars of *Brassica napus, Brassica rapa* subsp. *oleifera* (syn. *B. campestris*), or *Brassica juncea*) is an important crop used to produce canola oil. In brief, Monsanto sued Schmeiser for collecting and sowing patented Roundup™-Ready canola seeds without a licence.[28]

Schmeiser claimed that in 1996, he had collected seeds from canola plants that had invaded areas around power lines and in ditches next to his land. He had sprayed these areas with Roundup™ herbicide. Noticing that around 60% of the canola plants survived the spraying, he collected seeds and conducted a field test in 1997 to document herbicide resistance more accurately. He subsequently collected seeds from these resistant canola plants, stored them over winter, and seeded a 1,000-acre field with them the next year. After Monsanto found out about Schmeiser's activities, they sued him for infringement of patent rights. They tested canola plants from Schmeiser's 1,000-acre field for the tell-tale herbicide-resistance protein, finding 95–98% of the plants were Roundup™-Ready Canola, confirming presence of the patented gene.

The Canadian Supreme Court heard protracted appeals of 'David vs Goliath', ruling five to four in favour of Monsanto.[28] The court determined that Schmeiser had infringed on Monsanto's patent right by knowingly planting Roundup™-Ready canola seeds. However, since he made no profit because of the 'invention' (i.e. the genetic modification of the crop), Schmeiser did not have to pay damages or Monsanto's very high legal fees. It is important to recognize that the court noted that Monsanto's patent was on the genetically modified genes and cells, not the plants themselves. This case set a precedent that patenting and subsequent licensing of GM crops were legally justified. And the presentation of molecular evidence (i.e. determination of the presence of the inserted resistance gene and its protein product) was acceptable to the court.

But more recently, the tables have turned as organic farmers are increasingly playing the plaintiff and taking the biotech companies to court. In essence, the farmers claim that the biotech companies are like trespassers on their land when their GM seed contaminates their crop. In court, the farmers can use the results of DNA technology tests to prove the presence of the *CP4 EPSPS* resistance gene or other 'foreign' genes that biotech companies have inserted into plants found growing on the farmer's land, or in hybrid seed produced from GM pollen fertilizing their organic crop.

Consumers are fighting back, too. Consider the 'taco' case. In 2002, James Byron Moran, the senior judge for the US District Court for the Northern District of Illinois, heard a consolidated class action suit on behalf of corn farmers against defendants Aventis Crop Science and Garst Seed Company.[29, 30] The plaintiffs claimed that the defendants had contaminated the entire US corn supply. This contamination occurred because the biotech company had engineered seed of yellow corn 'StarLink™' to express the Cry9C gene and produce a chemical toxic to insects. While StarLink was licensed for use on crops destined for animal feed, ethanol production, and seed increase on 2.5 million acres, it was not licensed for the cultivation of corn destined for human consumption.

Unfortunately, and perhaps as expected, pollen from Star-Link corn had drifted into fields of corn varieties without the

Cry9C gene. In addition, corn seeds from StarLink fields got into the food supply as it commingled with corn from other fields from hundreds of farms during storage and shipment. Testing revealed that the Cry9C gene was in many food products, including taco shells. Although the court found the case above in favour of the defendant, Aventis later settled for $110 million, and they withdrew StarLink corn from market in 2000. Many US food producers stopped using US corn, and South Korea, Japan, and other countries stopped or limited imports of US corn.

A similar case involved contamination of the 2005 rice supply because of an unapproved contamination from GM Liberty Link Rice (LLRice) grown in field tests.[31] Scientists at the biotech company Bayer CropScience genetically engineered LLRice to be resistant to the chemical glufosinate, the active ingredient in Liberty herbicide used to kill weeds in rice, soybean, cotton (*Gossypium* spp.), and canola fields. Again, many countries imposed emergency bans on the import of American rice or required testing and certification procedures be put in place. Between 2005 and 2008, exports of US rice decreased by 622,972 metric tonnes. In this case, in 2011 Bayer CropScience settled the class action lawsuit of 11,300 rice producers for $750 million.

What can a farmer do if they suspect contamination in their crop from a nearby GM crop? Besides straightforward herbicide spray tests, there are now commercial companies offering services using PCR-based tests to detect the genes and, indirectly, the proteins in GM crops. Broad-spectrum tests can determine if bacterial or viral elements are present in the DNA (as they will be in a GM plant). The biotech company inserts these elements to regulate trait expression of the relevant gene (e.g. to regulate expression of the gene conferring herbicide resistance). By contrast, event-specific or construct-specific PCR tests can detect a specific engineered gene. Scientists are developing even more specific tests than PCR based on so-called next-generation sequencing (NGS, also called massively parallel sequencing: MPS) in maize and cotton. The NGS-based tests offer promise for more rapid, accurate, and precise forensic testing of GM traits than is possible using conventional PCR.

Genetically modified crops are getting harder to detect. Scientists can now carry out small non-GM genetic edits not involving insertion of bacterial or viral genes using procedures such as CRISPR, which is derived from studies of bacterial DNA fragments. In short, bacteria have repetitive 'spacer' DNA sequences that match bacteriophage (viral) sequences, and they are used to counter attacking viruses. Scientists can adapt this mechanism to make very small edits to a crop's genome at precise locations to confer new traits. Gene-edited soybean and lettuce (*Lactuca sativa*) are already in production, with scientists developing new varieties of several other crops including potatoes (*Solanum tuberosum*), strawberries (*Fragaria* x *ananassa*), and bananas (*Musa* spp.).[32] These edits are much smaller than the conventional insertion of foreign DNA segments in older GM crops. Moreover, the small edits may be similar and essentially indistinguishable from natural mutations.

The similarity of genetically modified crops to those arising from natural mutations has led to a conundrum for food testing laboratories and the lawyers waiting to prosecute cases of patent infringement or crop contamination. While many countries, including Brazil, Argentina, Australia, and the United States, have decided not to regulate crops produced from such small edits, in July 2018, the European Union ruled a requirement for mandatory testing. Gene editing legal expert Martin Wasmer of Leibniz University, Germany expressed concern that crops produced in this way may be impossible to distinguish from naturally occurring organisms, making enforcement impossible.[33]

The new molecular technology of NGS may be helpful here as it allows very rapid sequencing of nucleotides from DNA in a single run. Additionally, NGS may be particularly useful for detecting small non-GM genetic edits. Advancements in NGS testing may be especially important in these food security issues and are driven in part by the need for the food industry to self-police authenticity of food products, hoping to stave off lawsuits.

NGS technology is also poised to provide a boom in the extremely difficult business of identification of fungal taxa. Unless a fungus is fruiting (producing mushrooms and spores), the fungal body has too few features to allow discrimination among different fungi except among major groups (e.g. yeasts compared with moulds). However, there is often the need to identify fungi in cases related to environmental toxicity, such as *Aspergillus* mould in buildings, *Penicillium*, *Debaryomyces*, and *Wickerhamomyces* on contaminated food, and even suites of fungi colonizing human cadavers to allow for estimates of PMI.

Scientists can accomplish DNA barcoding for fungal identification based upon ITS genes along with accurate reference databases. Rolf Nilsson and colleagues have developed the UNITE database that has been available since 2013 for the molecular identification of fungi.[34, 35] This database has an incredible 2,688,805 ITS sequences, allowing identification of 457,996 hypothesized fungal species as of December 2020. NGS technology promises to make the identification of fungi in forensic cases even better and faster and even allows the identification of mixtures of species.

* * * * *

The use of DNA in botanical forensics will only continue to grow with methodological advances. An exciting new development is the forensic application of eDNA (environmental DNA) methods. Living organisms shed DNA into the environment with which they interact. NGS and DNA metabarcoding allows identification of animals, plants (including fragments, pollen grains, and diatoms), fungi, and the soil microflora from mixtures of their DNA in soil, water, and air samples. Described as a 'game changer', eDNA sampling is fast, inexpensive, and highly sensitive. It has allowed the tracking and monitoring of rare species and invasive species. Forensically, when samples are collected from a suspect or associated with some evidence (e.g. soil on a suspect's car tyres or clothing) then it may be possible for the identity and provenance of the organisms in the sample to be determined.[36, 37] For example, investigators used DNA metabarcoding to identify several invasive plants species in a proof-of-concept study of eDNA in freshwater samples from a Canadian lake and a river. The plants identified were known to be available from local plant nurseries and aquarium supplies and to have escaped into the wild.[38]

In a pioneering case, eDNA evidence was accepted by the court to support the occurrence of invasive, non-native Asian carp in the man-made canals of the Chicago Area Waterway System (CAWS) following their migration up the Mississippi River. Plaintiffs were seeking to compel the US Army Corps to close the

Chicago locks as the migration of the carp through the CAWS would cause an environmental threat to Lake Michigan.[39,40] eDNA cases involving plants are sure to follow.

Another exciting new development using DNA analysis is the application of methods to analyse the plant microbiome.[41] Plant fragments can be important sources of forensic evidence, as we have seen, but they can be very difficult to identify, even for a trained botanist. However, the community of microbes colonizing and inhabiting leaf, stem, flower, and fruit surfaces (the phyllosphere) is unique to each plant species, the location in which the plant is growing, and the season when the samples are collected. Retrieval of DNA from the phyllosphere and identification of the taxa comprising the microbiome through use of high-throughput DNA sequencing can provide a powerful additional tool for forensic investigators. This forensic application of microbial ecology has its origin in the examination of the microbiome inhabiting cadavers and the soil over clandestine graves.[37] Much like the use of eDNA described above, the utility and application of the plant microbiome approach for forensics is in its infancy.

As new sources of evidence, eDNA information and analyses of the plant microbiome will need to meet *Frye* and *Daubert* challenges to be accepted in court. And there are challenges to overcome in establishing consensus in methodologies and the standardization of statistical analyses.[42]

In the next chapter, we move from a molecular to a chemical focus as we look into the forensic need to identify plant poisons. The popularity of plant poisons as a murder weapon of choice may have diminished somewhat following the availability of advanced chemical methods of identification and testing. In medieval times, purveyors of potions were often also professional poisoners. Nevertheless, plant poisons are still sometimes deliberately used to 'knock off' an unwanted individual; in these cases, plant sleuths need to identify the poison and present forensic evidence in court.

A FORENSIC PHARMACOPOEIA

> I have heard that one should die in silence. Keep quiet and be brave.
>
> Socrates, after drinking a poisonous cup of hemlock[1]

After being given the death penalty in 399 BCE, Socrates famously chose to drink deadly hemlock tea as his method of execution. With a brew made from as few as six to eight leaves of *Conium maculatum* (poison hemlock), Socrates consigned himself to a numbing death as Plato reports that his muscles became paralysed from the feet up to the heart because of the toxic effects of coniine, a plant alkaloid. Plato's poetic account of Socrates' death[1] is likely sanitized and fictional, as typical symptoms of hemlock poisoning include vomiting, slurred speech, convulsions, and finally asphyxiation as the respiratory muscles become paralysed.[2]

People have used plants to poison victims since time immemorial; they have also caused accidental poisoning, and depressed individuals have used them in suicides. When a poisoning occurs, there's a need to identify the plant either to save the victim or to help find the perpetrator.

In this chapter, we will look at the different types of toxins from plants and fungi, and how they have caused harm to humans, accidentally and deliberately. In keeping with the rest of this book, the focus is on cases of plant poisoning that involve forensic and legal investigation at some level. These investigations involve identifying the toxin and determining culpability and blame. I will not dwell here on the methods of chemical detection of plant toxins, as these have been described in other books.[3] Broadly defined, a poison is any substance that can cause death, injury, or harm, and a toxin is a poison that plants, animals, algae, viruses, bacteria, and other organisms in nature produce. Venoms are poisons that an animal delivers to the affected organism through biting (e.g. snakes) or stinging (e.g. wasps). There is some blurring of the lines between these definitions. For example, technically, we may consider the painful cocktail of chemical toxins that are injected into the skin through the hyperdermic-like trichomes on the leaves of stinging nettles (*Urtica dioica*) to be a venom.

The history of poisons parallels that of medicinal plants, since substances beneficial at a low dose often become poisonous at a high dose. An ancient Egyptian pharmacopoeia with more than 900 medical prescriptions contained several derived from highly poisonous plants including aconitine (from *Aconitum napellus*), atropine (from *Atropa belladonna*), opium (from *Papaver somniferum*), and physostigmine (from *Physostigma venenosum*).[4]

Belladonna (*Atropa belladonna*), also called deadly nightshade, is a good example to start with as it has beneficial properties, but has also caused accidental poisonings, and has been used by murderers.

Belladonna was used by women to enhance their beauty. During the Renaissance, Venetian women used eye drops prepared from belladonna at low doses to dilute their pupils. The very name of the plant means 'beautiful woman' in Italian. Belladonna has also been used as a dietary supplement and alternative medicine. However, belladonna plants are extremely toxic, containing tropane alkaloids (principally atropine, chemically DL-hyoscyamine) that cause many medical problems, and can result sometimes in death. Indeed, belladonna was used in ancient times in making poison arrows, and scholars believe that Livia used it to kill her husband, the Roman Emperor Augustus. The Parthians in Asia Minor used the closely related devil's trumpets or jimsonweed (*Datura stramonium*) to poison Mark Antony's troops in 36 BCE.[5]

Atropa and *Datura* are in the Solanaceae plant family that includes the more innocuous tomatoes (*Solanum lycoperiscum*), potatoes (*S. tuberosum*), and aubergine (*S. melongena*), as well as the toxic nightshades, black nightshade (*Solanum nigrum*) and woody nightshade (*S. dulcamara*). Scholars have long known of the culinary, medicinal, and toxic properties of these plants. The Greek philosopher Theophrastus wrote in his *Enquiry Into Plants* that 'one is edible and like a cultivated plant . . ., one is said to induce sleep, the other cause madness, or, if it is administered in a larger dose, death'.[6]

Poisoners may consider atropine from belladonna an attractive choice for murdering someone because it is quick to act. It spreads throughout the body without inflaming any internal organs, and it rapidly breaks down, disappearing in buried corpses. Perhaps these are the qualities that led Paul Agutter to use atropine in the attempted murder of his wife Alexandra in 1994.

Agutter was a biology lecturer at Napier University, Edinburgh, Scotland, and part of a biomedical research group specializing in toxicology. He had access to atropine sulphate, and knew the dose he would need to murder his wife. He also knew that he had to disguise its bitter taste and administered the poison by spiking his wife's gin and tonic. To carry out the 'perfect murder', he spiked several one-litre bottles of tonic water with a low dose of atropine, placing them back on the shelves of his local Safeway supermarket. He hoped that others would purchase and drink this tainted tonic water, causing them to get sick but not die. The police would then consider his wife's death part of a more widespread poisoning spree carried out by an unknown person.

Agutter's plan didn't go as he hoped when he put it into action on Saturday evening, 28 August 1994. His wife tasted the bitterness of the atropine so only drank about half of her gin and tonic. This amount gave her a serious, but non-lethal dose of atropine. To make things worse for himself, he didn't dispose of the evidence. The paramedic who took Mrs Agutter to hospital suspected that someone had poisoned her, and so they collected the tonic bottle and the remains of her drink for analysis.

A witness had also spotted Paul Agutter putting the spiked tonic water bottles that he had bought earlier back onto the shelf in the supermarket. Also, by sheer bad luck, one of the spiked bottles was purchased by Geoffrey Sharwood-Smith, a local consultant anaesthetist who was familiar with the symptoms of atropine poisoning. After Sharwood-Smith's wife and son became sick after drinking some of the tonic water, he rushed them to hospital and informed the police of his suspicions.

Alexandra Agutter was one of eight victims of atropine poisoning admitted to the local hospital by the end of August 1994. They had all drunk tonic water purchased from the same store. This 'tonic water scare' led to a nationwide recall of the Safeway brand of tonic water.

The evidence caught up with Agutter, and police arrested him and charged him with the attempted murder of his wife. At trial, forensic scientist Howard Oakley testified that the bottle of tonic used to make Mrs Agutter's gin and tonic had 292 mg of atropine, and the other bottles from the supermarket had 11–74 mg per litre. Mrs Agutter's dose would have been fatal if she had drunk all of her gin and tonic. Judge Lord Morrison sentenced Agutter to 12 years in jail, saying, 'This was an evil and cunningly devised crime.'[7,8]

Accidental poisonings from belladonna also occur. Just a few years ago, family members brought a 49-year-old woman into the emergency room in a Turkish hospital. She was suffering from fatigue and delirium that had started three to four hours after ingesting 'forest fruits'. Shortly after admission to the hospital, she lost consciousness. She was presenting symptoms indicating that the neurotransmitter acetylcholine was being blocked, including irregular breathing and elevated heart rate, and her pupils were dilated. The physicians assessed these symptoms as indicative of potentially severe brain injury.

Her relatives brought in samples of the plant that she had been eating. The plant had the characteristic shiny black berries and leaf and flower features of *Atropa belladonna*. The impact of alkaloids in the belladonna (L-atropine, DL-hyoscyamine, and hyoscine) on the central nervous system caused the symptoms that the doctors observed. She was treated with mechanical ventilation, gastric irrigation, the administration of activated charcoal via a tube to bind and deactivate the toxin, and sedation. The woman was very lucky. Her symptoms were gone after 24 hours in the intensive care unit, following which she was transferred to a general ward before being discharged from hospital.[9]

* * * * *

Forensic toxicology is the study and application of toxicology, and analytical chemistry and pharmacology, to assist medical and legal investigations of deaths and drug use. This field covers many toxins, ranging from the radioactive polonium 210, which Russian agents used in 2006 to poison former KGB agent Alexander Litvinenko in London,[10] to the nerve agent VX, which two Vietnamese women sprayed in the face of Kim Jong Nam, with fatal results, at Kuala Lumpur International Airport in 2017, on the orders of his half-brother, North Korea's leader, Kim Jong Un.[11] However, plants have perhaps been the most popular deadly poisons, especially before the development of synthetic chemical toxins.

Widely available plant products can be toxic at high concentrations. Something wasn't right when the seemingly healthy Gustave Fougnies died during dinner on 20 November 1850. His hosts claimed he died of a stroke, but he had actually been poisoned with nicotine extracted from tobacco plants (*Nicotiana tabacum*). Tobacco, of course, is a legal plant drug used by millions of people worldwide.

A single man, Fougnies had inherited money and property. But when he announced plans to marry, his sister Lydia Fougnies du Bois and her husband Hyppolyte Visart de Bocarmé (the Count and Countess of Bocarmé) realized they would lose their inheritance from Gustave. An amateur chemist, Bocarmé purchased tobacco under a false name, stockpiling the leaves in his basement over the summer of 1850. He worked with local chemists to extract nicotine from the leaves, testing the poison in experiments on cats and dogs. Nicotine is a plant alkaloid naturally occurring at concentrations of up to 7.5% in the leaves of tobacco plants. At high doses nicotine poisoning can lead to respiratory paralysis and death.

The Count and Countess of Bocarmé invited 32-year-old Fougnies to their Château de Bitremont in Wallonia, Belgium for the dinner during which he died. Their servants were suspicious, since Fougnies had bruising and caustic burns around his face, and so they contacted the authorities. How could the suspected poisoning be proven?

Thankfully, Belgian chemist Jean Servais Stas developed a method to isolate alkaloids from preserved bodily tissues and test for nicotine. He used acetic acid solution in warm ethanol to extract nicotine from Fougnies's preserved organs. Stas's method was the first toxicological isolation of a plant alkaloid. Police charged Bocarmé, the court found him guilty of murdering Fougnies, and he was executed. The German chemist Frederick Otto later modified Stas's test into what became the Stas–Otto method, enabling the isolation of many plant alkaloid poisons. This case and development of the Stas–Otto test spurred the development of forensic toxicology.[12]

Despite the large number of plant poisons, only about 40–50 of the approximately 16,000 or more genera of flowering plants around the world are responsible for most cases of plant poisoning. Patients usually recover if they can get medical attention quickly enough. Biochemists place the plant toxins responsible into several important chemical groups: alkaloids (including tropane alkaloids), glycosides, plant lectins, amatoxins, and cyanotoxins.[13] We'll look at these in turn.

There are more than 5,000 naturally occurring plant alkaloids, including caffeine (from tea (*Camellia sinensis*) and coffee (*Coffea arabica* and *C. canephora*)), cocaine (from coca *Erythroxylum coca*), colchicine (from the crocus *Colchicum* spp.), morphine (from the opium poppy), and nicotine (from tobacco). Biochemists characterize plant alkaloids by the possession of basic nitrogen atoms in addition to their carbon and hydrogen. These compounds may contain sulphur and oxygen, or both, and sometimes also bromine, chlorine, or phosphorus.

The tropane alkaloids are a class of alkaloids containing a nitrogenous structure in their molecule, and they include several secondary compounds found in members of the Solanaceae plant family. We have already noted that atropine occurs in deadly nightshade, and we find hyoscyamine in henbane (*Hyoscyamus niger*), mandrake (*Mandragora officinarum*), and the sorcerer's tree (*Latua pubiflora*). Additionally, scopolamine occurs in henbane and jimsonweed, and strychnine occurs in the strychnine tree (*Strychnos nux-vomica*).

Glycosides are molecules in which a sugar is covalently bonded to another functional chemical group. These plant toxins include digoxin (digitalin, from the leaves of foxgloves, *Digitalis purpurea*) and oleandrin (from oleander, *Nerium oleander*). Digoxin has featured in many deaths, including the poisoning of Italian warlord and autocrat Cangrande della Scala in 1329[14] and in somewhere around 400 deaths by the 'Angel of Death' serial killer Chris Cullen in the 1980s and 1990s.[15]

Doctors use digoxin at low doses for patients with congestive heart disease. Nevertheless, Cullen over-prescribed and unnecessarily prescribed digoxin to kill patients while employed as a nurse at 16 hospitals in New Jersey. He was finally brought to justice through digital forensics after detectives investigated computer records from hospital drug dispensaries. He was subsequently sentenced to multiple life sentences for his crimes. In response to Cullen's ability to move from one hospital to another when his behaviour raised suspicions, the state of New Jersey passed the so-called Cullen Act which requires healthcare facilities to report professional misconduct of their staff.[16]

Plant lectins are large carbohydrate-rich proteins consisting of two protein chains and include the highly toxic ricin from castor bean, and abrin from jequirity bean (*Abrus precatorius*). Ricin has been involved in some high-profile murders, as we shall see. Lectins inactivate the subcellular organelles called ribosomes that synthesize proteins, causing death at very low doses.

Mushroom poisoning is well known, and fungi have their own suite of poisons, amatoxins. Unfortunately, few people are able to identify toxic mushrooms. Certain mushrooms carry a large variety of toxic amatoxins. These amatoxins usually become a health problem because of accidental ingestion of misidentified mushrooms, although there are cases of deliberate mushroom poisoning. There are several amatoxins, but they are chemically similar. These toxins can be deadly, as they impair protein synthesis, especially in the liver. They occur in several mushroom genera, including species of *Amanita*, such as the death cap mushroom (*Amanita phalloides*).[13, 2]

Cyanotoxins are produced by some algae and bacteria. For example, researchers found the deaths of 26 of 130 patients following kidney dialysis over a four-day period in February 1996 at a healthcare facility in Brazil to be due to algae and bacteria.

Investigation revealed that the source of water for the dialysis procedures was a reservoir contaminated with microcystins, a cyanotoxin. Although not usually the focus of criminal activities, some freshwater cyanobacteria and marine dinoflagellates produce cyanotoxins during freshwater 'blooms' (periods of high growth rates). In salt water, dinoflagellate 'red tide' blooms can similarly produce deadly neurotoxins (dinotoxins) that can lead to fish kills and human health problems. Saxitoxin, a potent neurotoxin, is produced by toxic dinoflagellate algae, and reaches high levels during periods of algal blooms. The neurotoxin is taken up by shellfish which when eaten by humans can develop into paralytic shellfish poisoning (PSP). In 2020, the Alaska state medical examiner's office reported that an individual had died following PSP that developed after consuming blue mussels collected from Dutch Harbor beach.[17,18]

* * * * *

There are written reports of plant toxins going back to ancient Chinese medicinal records. Shen Nung, second Celestial Emperor and regarded as the father of Chinese medicine, recorded medicinal and poisonous properties of 1,000 herbs in his treatise *On Herbal Medical Experiment Poisons*. He allegedly tasted 365 herbs, discovering that tea (*Camellia sinensis*) was an antidote to 70 poisonous plants. Unfortunately, so the legend goes, after eating the yellow flower of a weed (likely the deadly *Gelsemium elegans*), his intestines ruptured. He died before he could swallow the tea he knew to be an antidote.

Notwithstanding Shen Nung's advice, be careful of which tea you drink. Poisons of forensic interest date back at least to 13th-century China. The government required local officials to investigate and produce written reports of deaths. In response, Chinese physician, judge, and forensic scientist Sung Tz'u published a manual titled *The Washing Away of Wrongs* in 1247. This manual listed the procedures that they required officials to follow in their investigations. The manual also included listings of the symptomology and descriptions of two toxic plants, rat-grass, or Japanese star anise (*Illicium religiosum*), and gelsemium root (*Gelsemium elegans*).[19,20]

Illicium religiosum and *G. elegans* are interesting choices for Sung Tz'u to focus on in his manual. *Illicium religiosum* (now named *I. anisatum*) is a well-known, highly poisonous evergreen shrub in Asia in the Shishandraceae plant family that also includes the non-toxic Chinese star anise (*I. verum*) used in cooking. There are several toxic chemicals present in *I. religiosum*, including the neurotoxin anisatin. When ingested even in small amounts, anisatin quickly causes epilepsy, hallucinations, convulsions, and ultimately death. When dried, the leaves of these two species are impossible to distinguish except through microscopic examination. Because of its similarity with *I. verum*, leaves of *I. religiosum* have been accidentally mixed into Chinese star anise tea, leading to poisonings and product recalls.

Gelsemium elegans, also called heartbreak grass (because of its use in suicides, although it is not a grass), is also well known for its highly toxic properties, as Shen Nung unfortunately discovered. This plant is a twining vine that grows to 12 m tall. It has a dense spike of clustered yellowish-white flowers and dark-green, blackish-coloured leaves. Gelsemine is the most toxic of 121 strychnine-related alkaloids that biochemists have identified from this plant. This chemical causes strong neurological and respiratory effects. Similarly, the New World yellow jessamine (*G. sempervirens*) is highly toxic. In extreme cases, this plant causes patients to die within 30 minutes of ingestion.

Fortunately, there have been relatively few poisonings because of *G. elegans*.[21] However, a 2017 report of accidental ingestion of the plant described a 26-year-old woman brought to a hospital emergency room after being found unconscious in her bedroom at her home in Teochow, China. On admission, she exhibited euphoria and childish behaviours and was unresponsive to verbal commands or pain. Doctors immediately placed her on a mechanical ventilator. Gradually, her health improved, but she could not be discharged for 36 days. Eight months later, she still exhibited some signs of neurological impairment. Investigators found a bottle of herbal broth on her bedside table, but it wasn't until three and a half months after the incident that the woman could confirm ingesting the broth even though she knew it would be toxic. To the doctors, the herb in the broth looked like *G. elegans*, and scientists at the China National Analytical Center confirmed its identity.[21]

In the past, plant poisoning was a common murder 'weapon' of choice. The Roman Emperor Nero used amygdalin, a cyanogenic glycoside, to dispose of unwanted family members. Peach (*Prunus persica*) seeds contain amygdalin, and ancient Egyptians and Romans used them for executions to facilitate the 'penalty of the peach'. In ancient Rome, the work of poisoners was countered by professional food and drink tasters, or 'degustators'. We still refer to the careful, appreciative testing of small portions of food as degustation.

Furthermore, the custom of clinking glasses supposedly arose as a way to spill some of your wine into your enemies' cup to provide some assurance that you'd both drink the same safe (or poisoned) beverage. Some scholars suspect that Alexander the Great's jealous wife, Roxana, poisoned him with strychnine from the tree *Strychnos nux-vomica*. Certainly, the symptoms of convulsions and hallucinations ancient scholars report him to have suffered as he died are consistent with this poison. And Roxana had access to *S. nux-vomica* and knowledge of its properties. Two years prior, she had visited a place in India where *S. nux-vomica* grows naturally and where healers were using it at low dosages to induce hallucinations.[4]

In the Middle Ages, poisoning was an important way of disposing of enemies standing in the way of personal and political gain. Catherine de' Medici, the 16th-century Queen of France, was known as the 'Queen-Poisoner', as she experimented with various toxins on the poor and sick. During the reign of King Louis XIV of France, '*l'affaire des poisons*' was a major scandal implicating many prominent members of court, including the king's mistress Madame de Montespan. Fortune-teller and commissioned poisoner Catherine Monvoisin supplied various types of powders ('*poudres de succession*'—hereditary arsenic-based powders, as they were euphemistically called) to members of the king's court.

If a member of the aristocracy wanted to get rid of a relative, five or six drops of the insidious 'Aqua Tofana', invented and sold by 17th-century Italian poisoner Giulia Tofana, was enough to kill. This water-like, odourless, and colourless concoction was a mixture of sulphuric acid, lead acetate, possibly arsenic, and a poisonous plant extract of either belladonna or ivy-leaved toadflax (*Linaria cymbalariae*, currently known as *Cymbalaria muralis*). Eventually, after supposedly 600 deaths, authorities in Rome sentenced Tofana to death and burned her at the stake in July 1659.

* * * * *

Nowadays, deliberate poisonings using plant toxins are less common than in the past. But they attract a lot of attention when they occur, especially in the Western developed world. These poisonings often involve chemically synthesized plant toxins rather than direct extracts from plant tissues. For example, serial poisoner Dr Harold Shipman killed somewhere around 200 patients in the 1980s and 1990s with morphine and diamorphine (heroin), which he obtained under prescriptions from pharmacies rather than by direct extraction from the opium poppy. Similarly, the digoxin used by Chris Cullen mentioned earlier was a synthetic pharmaceutical product and not a direct extract of foxglove.

As already mentioned, strychnine has been a frequently used plant-based poison from Ancient Greek times through to the present day. Risus sardonicus or rictus grin is the unmistakable, grotesque smile characteristic of poisoning by this neurotoxin. Under the category of 'did he really think he would get away with it' is an unusual modern-day case. In this case, in 1990 police accused 46-year-old former professional American football player John Morency of poisoning his wife, 30-year-old Sue Morency, with strychnine in their Kensington Park Villas, San Diego home. Strychnine is an alkaloid extracted from the seeds of the shrubs *Strychnos nux-vomica* and *S. ignatii* in the Loganiaceae plant family found in India and the Philippines. Homeowners and pest-control professionals formerly used it as a rodenticide until its extreme toxicity was recognized to be a danger to children. In the form of a bitter-tasting but odourless crystalline powder, poisoners can mix strychnine into drinks and food. John Morency dissolved it into his wife's tequila.

The couple were having marital problems. Friends claimed that Sue Morency was planning to leave her husband. It was then perhaps no coincidence that John Morency called 911 for emergency help on the evening of 21 July 1990, claiming that his wife was suffering from food poisoning. Two days earlier, her husband had doubled her life insurance policy. When paramedics arrived, Morency tried to send them away, downplaying his wife's sickness and saying that they were planning on leaving for a holiday to Cabo San Lucas, Mexico the next morning. Hearing her screams, the paramedics ignored him and found her nude, convulsing on her bed, and crying out in pain. Shortly afterwards, she went into cardiac arrest while in the ambulance en route to the Sharp Memorial hospital. In the hospital, doctors placed her on life support, but she died a few days later on 5 August without regaining consciousness.

The clinical symptoms of strychnine poisoning are fairly clear because the poison causes simultaneous muscle contractions to the point that the patient suffocates. As the back and neck muscles are particularly powerful, the body arches backwards, the upper limbs are flexed onto the chest, and the jaws are totally closed, causing the person to exhibit the rictus grin. Medical professionals refer to this agonizing spasming of the muscles as opisthotonos (see Figure 11). There is no specific antidote for strychnine poisoning, and the hospital doctors were initially unsure of what was causing Sue Morency's symptoms. With no antidote, she died an agonizing death. An autopsy revealed that she had four times the lethal dose of strychnine in her body. At trial, the jury convicted John Morency, and the judge sentenced him to life imprisonment.[22]

There are several other famous cases of strychnine poisoning. I have already mentioned the poisoning of Alexander the Great through the spiking of his wine in 323 BC. In 1905, Jane Stanford,

Fig. 11. Opisthotonos in a patient illustrating the classic symptoms of tetanus and strychnine poisoning.

co-founder of Stanford University, drank from a bottle of bicarbonate of soda poisoned with strychnine. Shortly before being seized by a fatal opisthotonos spasm, she exclaimed, 'My jaws are stiff. This is a horrible death to die.'[23] And it is alleged that Delta blues legend Robert Johnson died in 1938 aged only 27 after his whisky was laced with strychnine. The circumstances of his death are unclear, much like the mysteries surrounding his life. Some say a jealous husband poisoned him, other accounts suggest he drank bad whisky, and still others that he died of syphilis.

Strychnine is not necessarily the go-to plant poison of choice for assassinations. That honour is perhaps reserved for the plant toxin ricin. Ricin and the related abrin from *Abrus precatorius* are toxalbumins, or glycoprotein lectins. Botanists find ricin in the seed endosperm (nutritive tissue) of the castor bean or castor oil plant (*Ricinus communis*), which is a member of the spurge family Euphorbiaceae. This plant is easy to obtain, and the toxin is easy to extract at low doses, but fortunately it is difficult to get in the high-quality, high-concentration amounts required to be lethal.

Specifically, the bean pulp can produce ricin following sulphuric acid extraction of the castor oil (which doesn't contain ricin). The pulp or mash contains approximately 5% ricin by weight. Poisoning with ricin can be as simple as through inhalation, ingestion, or injection. Symptoms appear within 12 hours of ingestion, or within 8 hours after inhalation of an aerosol containing ricin. The symptoms are initially non-specific, but they progress to multi-organ failure, cardiovascular collapse, and death. Ricin is fatal because it inhibits protein synthesis through inactivation of the ribosomes. As ricin is more lethal than cobra venom, the US government considers it so toxic that it is listed as a biological 'select agent' or toxin (BSAT). Just 280 µg of ricin, about 1/100th of a grain of rice, would most likely kill a 70 kg (154 lb) person—enough to do me in.[24]

Because of the ease of extraction and high toxicity, murderers have used ricin for assassination attempts. In October and November 2003 and in February 2004, an unknown person sent ricin powder in letters to Greenville, South Carolina, the White House, and US Senator Bill Frisk, respectively. Fortunately, alert clerks intercepted these letters in a mail-sorting facility before they reached their intended victims. The perpetrator signed threatening notes enclosed with the packages as 'Fallen Angel', although authorities never identified or arrested the individual responsible.[25]

In 2013, a well-known Hollywood movie star sent anonymous, threatening ricin-laced letters to then New York City mayor Michael Bloomberg, and US President Barack Obama. The actress was bizarrely attempting to frame her estranged husband. She was caught, pleaded guilty, and was sentenced to 18 years in jail, and her acting career was over.[26] Attempts to poison politicians continue with a report of a possible, but failed effort to send ricin powder to then President Trump in 2020.[27]

Ricin has been weaponized for bioterrorism. During the First and Second World Wars, the United Kingdom and the United States conducted studies on how to use ricin in bullets and bombs as a biological weapon, but never used it. Later, in the 1980s, Iran also tried to weaponize ricin. More recently, in 1999, authorities arrested an al-Qaeda terrorist for trying to blow up Los Angeles airport. At his trial, the defendant claimed that his group was planning to smear door handles with ricin as part of their terror campaign. Finally, police suspected a terrorist group arrested in London in 2003 of trying to extract ricin. Because of these threats, the international Biological Weapons Convention and Chemical Weapons Convention list ricin as a controlled substance, and ban its production, stockpiling, and transfer.[28,29,30]

Despite the international ban on use of ricin as a bioterrorism agent, the Soviet Union used it in state-sponsored assassination attempts, such as the 'umbrella assassination'. Bulgarian defector Georgi Markov was living in the United Kingdom in 1969 after escaping from his communist home country. Working as a BBC journalist, he was openly critical of the pro-Soviet Bulgarian regime. In September 1978, while waiting at a bus stop on Waterloo Bridge in London, Markov felt a prick on his leg. He turned to see a man bent down picking up an umbrella. Speaking with a foreign accent, the man apologized, hailed a taxi, and left. A few hours later, Markov developed a high temperature and was admitted to St James' Hospital, Balham. His symptoms rapidly worsened: he was running a high fever and had abdominal pains, diarrhoea, vomiting, and a high white blood cell count. The next day, he went into shock, and he died three days later.

An autopsy found a 1.55 mm wide metal pellet in Markov's thigh. This pellet had an x-shaped cavity and two 0.34 mm drilled holes that allowed the poison to be injected into the pellet. Covered in wax that melted at body temperature, the pellet released 500 10^{-3} ml of ricin. The pellet was an extremely sophisticated, precision-engineered poison-delivery device. Investigators determined that the man's umbrella had been adapted as a compressed air-gun with a trigger in the handle to inject the pellet containing ricin into Markov's leg.

This assassination wasn't the first time a murderer had used such a device to deliver ricin. Just two weeks before Markov's assassination, Vladimir Kostov, another Bulgarian exile, became ill from injection into his back of an identical pellet while he was travelling in the Paris metro. Fortunately for him, the wax around the pellet didn't melt, so very little of the ricin was released into his body. Doctors recovered the pellet and identified the ricin. Former KGB agents Oleg Kalugin and Oleg Gordievsky admitted Soviet complicity in planning these assassinations five years later. The identity of the actual assassin in both cases remains unknown, although scholars suspect it to be that of Francesco Gullino, a Dane of Italian origin. Gullino was working during this time as a Soviet-backed, Bulgarian spy referred to as 'Agent Piccadilly'. The whole truth remains a Cold War mystery.[31]

Another plant poison that has been favoured by the Kremlin is gelsemine. Alexander Perepilichnyy died while jogging near his home in London on 12 November 2012. He was a 44-year-old, otherwise healthy man. His death was initially not suspicious, and authorities thought it to be from an unfortunate heart attack. However, scientists with the Royal Botanic Gardens, Kew found high levels of toxins, possibly from 'heartbreak grass', the plant *Gelsemium elegans*, in Perepilichnyy's stomach following tests carried out three weeks later as part of a coroner's inquest.

Investigators revealed that he was a Russian whistleblower who had been subject to Kremlin death threats. He had spilled the beans on a fraud involving Russian tax officials, in which he stole £147 million from the Hermitage Capital hedge fund run by the Vladimir Putin critic, US-born financier Bill Bowder. A puncture mark on Perepilichnyy's neck supported the view that he was deliberately poisoned. Alternatively, he may have ingested the poison in the bowl of Russian sorrel soup he had for lunch if *G. elegans* powder or leaves had been mixed in with the sorrel (*Rumex acetosa*). Nevertheless, the cause of his death remains uncertain as it is reported that the UK government was reluctant to release information in their files on Perepilichnyy.[32]

* * * * *

Some poisons have unusual, non-toxic effects that perpetrators take advantage of. A 28-year-old professional woman had no memory of cashing her salary check, withdrawing money from bank cash machines, going to her home to get her jewellery, and giving it all to an elegantly dressed stranger. Yet she did these things. The episode began when she left her office in downtown Bogotá, Colombia at 11:00 a.m., and ended when she regained consciousness at home at 2:30 p.m. later the same day. A relative took her to the local hospital, where a urine test was returned positive for scopolamine and fenotiazine (a sedative).

'Million-dollar rides' from scopolamine (burundanga or 'devil's breath') poisoning are not uncommon in Colombia. Burundanga is made from extracts of seeds of so-called zombie plants *Datura stramonium* (jimsonweed), *Hyoscyamus albus* (white henbane), or *Brugmansia insignis* (borrachero tree). These plants are all in the Solanaceae plant family. Criminal burundanga poisoning has been occurring since the 1950s, and although reports of such cases are difficult to obtain, one hospital in Bogotá received 98 cases of burundanga intoxication between 1980 and 1981, and authorities recorded 102 cases between 1988 and 1989 in another. In one case, the criminal induced the intoxication by spraying the victim in the face. In another, a criminal robbed a 47-year-old Canadian man of $250 in cash and stole his mobile phone while he was in an amnestic state for 12 hours during a vacation in Bogotá. Although he was unaware of the robbery at the time, he was able to board a bus and return home despite his feelings of disorientation and confusion. More recently, since a 2016 case in which a woman's drink was spiked with a white powder, burundanga has frequently been used as a date-rape drug in Spain.[33,34,35]

Burundanga intoxication leads to confusion, euphoric behaviour, palpitations, dilated pupils, dry mouth, and, most importantly, submissive and obedient behaviour in the victim. Physicians refer to the phenomenon as transient global amnesia (TGA), in which victims or patients exhibit memory loss for a period of time while retaining their personal identity and the ability to perform everyday activities. TGA can occur following brain damage, but it can also be induced by drugs, such as scopolamine. Victims are unaware of what's going on, and there are usually no witnesses. The body rapidly detoxifies the drug, leaving little to no trace 12–24 hours later, making prosecutions difficult.

The scopolamine alkaloids implicated in burundanga have anticholinergic properties (including dry mouth, dilated pupils, blurred vision, and increased heart rate), as well as antiemetic (prevents nausea) and hallucinogenic ones. It acts by blocking acetylcholine in the nervous system. People variously use the plants containing these alkaloids for food, as ornaments (beads, necklaces), and for medicinal properties. As scop-dex (scopolamine mixed with dexedrine), it is used to reduce motion sickness in astronauts. But in larger doses, scopolamine is a toxin. Soviet agents used it in laced candy in an unsuccessful 1955 attempted incapacitation and abduction of Lisa Stein, an interviewer with the 'Radio in the America Sector' US propaganda station in West Germany.[36]

* * * * *

Most plant poisonings are accidental or sometimes self-induced suicide attempts. It can be difficult for investigators to tell accidental poisonings from suicides. In one case in the Netherlands, a 19-year-old woman was found dead in her hospital bed one morning. She had been admitted to a psychiatric hospital the night before because of a previous suicide attempt, and had no access to medication on her own, precluding an overdose. However, a search of her apartment found yew (*Taxus baccata*) leaves, a half-empty tea glass, and documents on the toxic effects of yew on her laptop. Although an autopsy was not performed and no yew leaves were found on her body or in her mouth or pharynx, her death was ruled a suicide due to ingestion of yew tea.[37]

Apart from the fleshy red aril surrounding the seed, all parts of yew trees contain taxine B, a highly toxic alkaloid. The toxicity of yew has been known since ancient times when the Celts poisoned arrows and spears with yew extract. Of 22 reported fatal suicides due to ingestion of yew between 1960 and 2016, 14 were due to ingestion of leaves, one of bark, three of the pulp, and four of a concoction or tea.[37] Ingestion of this plant and the release of taxine B into the body can lead to rapid cardio-respiratory failure and death within two hours. The presence of yew leaves around the mouth, on the tongue, and in the oesophagus and stomach usually leads investigators to suspect yew poisoning. Analysis of blood and urine can confirm the presence of taxine B or the related 3,4-dimethoxyphenol at toxic concentrations, although the taxine B compound is unstable and difficult to detect in post-mortem blood. A fatal dose can be a single cup or glass of yew-leaf tea or the consumption of a few leaves (about 50 g).[38]

Some plant toxins are used for both suicide and murder. *Cerbera odollam* (Apocynaceae), known as pong-pong, othalanga, or, as Yvan Gaillard and colleagues have more recently dubbed it, the 'suicide tree', is one example.[39] This plant is a medium-sized tree endemic to India and Southeast Asia. The tree produces very toxic seeds with a bitter taste that is disguised when mixed with spicy food. One seed kernel is enough to kill an adult. The seeds contain the heart toxin cerberin, a cardenolide or cardiac glycoside, similar to the well-known digoxin heart poison in foxglove. The toxin affects the heart's electrical activity, causing an erratic heartbeat, cardiac damage, and, if untreated, death. Ironically, part of the initial treatment for cerberin poisoning includes administration of the otherwise toxic plant alkaloid atropine, which we met earlier, along with insertion of a heart pacemaker.

A 2004 study reported that *Cerbera odollam* was involved in 50% of plant poisoning cases and 10% of all poisoning cases in Kerala, India. There were 500 fatal cases from 1989 to 1999. Three quarters of the victims were women, and Gaillard and colleagues from the Laboratory of Analytical Toxicology in France suspect that family members were using the poison to kill young wives who didn't meet the strict standards of behaviour of some Indian families. Poisoning due to *C. odollam* is very hard to trace, making it the perfect 'murder' tree. In this case, biochemical analyses used high-performance liquid chromatography (HPLC) with mass spectroscopy to chemically separate and measure the abundance of compounds. This analysis revealed several otherwise unnoticed homicides that investigators had previously thought to be suicides.[39,40]

Accidental plant poisonings occur for several tragic reasons. There are poisonings associated with herbal and traditional medicines, as well as adverse reactions to medicinal plants. Sometimes, users accidentally exceed the safe dose of an otherwise harmless plant. In other cases, the consumer has ignored the warnings that a plant is toxic. There are cases in which the plant is reputed to be nontoxic when it isn't, and cases in which a toxic plant is not listed on a label because of deliberate or accidental substitution or contamination.

Clinical symptomology can help identify the class of toxic chemicals involved in a poisoning, such as the rictus grin and opisthotonos already described for strychnine poisoning (Figure 11). Regardless, identification of the plant suspected in a plant poisoning case, whatever the cause, is critical and needs to be done fast. Accurate morphological identification of the plant is necessary and requires the involvement of expert botanists.

Sometimes, identification isn't too difficult. For example, in the atropine poisoning case described above, relatives brought in samples of the plant that the woman had ingested, and the doctors identified it as *Atropa belladonna* from the clinical symptoms and information in the literature. In suspected yew tree poisonings, a botanist can identify the poisonous plant as *Taxus baccata* from leaves found on or in the deceased. In other cases, there may be no incriminating plant remains to examine, or the victim ingested a powder or liquid extract from which a pathologist can carry out chemical blood or urine tests based on clinical symptomology. But while some toxins may quickly disappear from the body (e.g. scopolamine/burundanga), others, such as morphine, hang around and toxicologists can detect them in body tissues or hair of corpses even after they have been buried for some time.

* * * * *

Poisonings attributed to traditional or herbal remedies are unfortunately common. Over a five-year period (1991 to 1995), traditional remedies caused death in 206 cases, of which 43% included herbal materials.[41] Unfortunately, the composition of these traditional remedies is most often unknown because of the secretive nature of the mixtures. When tested, the plant toxins in these folk remedies are often found to be alkaloid-based. An African survey of traditional medicines[42] identified pyrrolizidines (a type of alkaloid) from ragwort (Senecio spp.) and rattlepod (Crotalaria spp.), atractyloside (a glycoside) from impila (Callilepis laureola), polycyclic hydrocarbons from the candelabra tree (Euphorbia ingens), methyl salicylate from the violet tree (Securidaca longepedunculata), and aloesin and aloeresin A (phenolics) from Cape aloe (Aloe ferox). At a high enough dose, these toxins can cause various maladies, ranging from liver and renal toxicity to nerve damage, that can lead to death. The kidneys are particularly susceptible to poisoning since they act to cleanse the blood and can accumulate toxins in the renal tissues. Traditional herbal medicines accounted for 30–35% of acute kidney problems in Africa, especially those containing aristolochic acid such as Mu Tong (Aristolochia manshuriensis).[43,44]

Prosecuting for suspected poisoning following ingestion of traditional medicines is difficult. In one case, a young man died following a shamanistic ceremony in the UK. He drank an ayahuasca infusion at the ceremony and died four days later. Ayahuasca is a traditional herbal brew that includes hallucinogenic alkaloids from bark of the vine *Banisteriopsis caapi* (ayahuasca, jugube, cappo, or yage) and the 'spirit molecule' the alkaloid dimethyltryptamine (DMT) from leaves of the shrub *Psychotria viridis* (chacruna). The shaman was initially arrested and charged with murder of the victim. However, microscopic examination of samples from the victim's colon found magic mushroom spores (*Psilocybe semilanceata*), *Cannabis* pollen, and opium poppy seeds. Apparently, the unfortunate young man had prepared himself for the ayahuasca ceremony with his own deadly concoction. The shaman's charges were reduced to possession of DMT, a Class A drug.[45,46]

I'd be extremely cautious about using traditional herbal medicines because of the problem of adulteration—the substitution of the intended herb with another toxic one. In the 1990s, more than 100 cases of very serious kidney problems arose because a Belgian clinic inadvertently switched one herbal ingredient for another in their 13-ingredient weight-loss formula. Physicians at the clinic inadvertently substituted the toxic perennial vine Guang Fang Ji (广防己, *Aristolochia fangchi*) for another vine, the intended Han Fang Ji (汉防己, *Stephania tetrandra*). The common names are very similar, differing only in the prefix Guang (广 'broadly' or 'generally') or Han (汉 'Chinese' or 'Sino-'), where Fang Ji (防己) means 'preventing snakebite for a hundred years'. But there was no excuse for this fatal mistake as these plants, known for their use in traditional Chinese medicine to treat snakebite, are in different plant families (Aristolochiaceae and Menispermaceae, respectively). A competent botanist would not mix them up, but by the time the roots are ground to a powder, telling them apart would likely be impossible without chemical testing. *Aristolochia fangchi* contains the plant alkaloid aristolochic acid, which is both toxic to the kidneys and a carcinogen that can lead to urothelial cancer. As a result, many European countries banned the medicinal use of all *Aristolochia* species and other related plants (e.g. *Akebia*, *Asarum*, and *Clematis*), and the US Food and Drug Administration discouraged the manufacture of products containing aristolochic acid.[47]

* * * * *

I must mention one of the most fascinating mysteries in forensic toxicology: the death of 22-year-old Christopher Johnson McCandless. Our understanding of this case is due to the persistent investigations of American writer and journalist Jon Krakauer, who published a biography of McCandless, *Into The Wild*, in 1996 that was also made into a 2007 movie.[48]

In April 1992, McCandless hitchhiked to Alaska, and started hiking along the Stampede Trail into the Denali Park wilderness area north of Mt McKinley. He had given up all his life possessions, including his $25,000 life savings and his car. Four months later, he was dead.

McCandless was from an affluent Washington DC family. He graduated with double honours in history and anthropology from Emory University, and he was an elite cross-country athlete. At the time of starting his journey, he hadn't spoken to his family in two years. He took only a bag of rice as provisions and a 0.22 calibre hunting rifle with him into the wilderness, explaining to local electrician Jim Gallien, who gave him a ride out from Fairbanks, that he was going to 'live off the land'. That plan didn't work out, as his diary reveals that he soon ran out of food and had trouble catching wild game and foraging for wild plants.

When some hunters found McCandless's decomposed and emaciated body on 6 September 1992, he weighed only 67 lbs and appeared to have starved to death. Diary entries reveal some of the problems he had. He tried to hike back out of the wilderness but could not cross a swollen river. He was fully aware of his predicament. As he starved, he wrote in his diary on 30 July 1992, 'Extremely weak, fault of pot[ato] seed. Much trouble to stand up, starving. Great jeopardy.'[48]

McCandless's diary and photos revealed that he had collected and eaten seeds (beans) of the Eskimo potato plant or Alpine sweetvetch, *Hedysarum alpinum*. This plant, in the Fabaceae or legume plant family, has nutritious rhizomes that native Alaskan people cook and eat. He was using an ethnobotany book of the Dena'ina Indigenous People from south-central Alaska to identify the plants he found. He correctly identified *H. alpinum* and had been eating the rhizomes early in the season, when they are edible. But later in the season, the rhizomes become tough and unpalatable, and McCandless, in desperation, switched to eating the seeds. Investigating McCandless's death, Jon Krakauer initially thought the seeds contained swainsonine, a highly toxic alkaloid present in the related locoweeds (*Astragalus* spp.). Locoweeds cause large economic losses in the livestock industry as they are common range plants readily eaten by cattle. Krakauer thought for several years it was this toxin that killed McCandless.

However, chemical analysis in 2007 showed that alkaloids are not present in *H. alpinum* seeds. Rather, after more investigation, in 2013 Krakauer and colleagues determined that a relatively unknown non-protein amino acid, L-canavanine, is present in the seeds of *H. alpinum* at concentrations of 1.2% by weight. L-canavanine is an analogue of the essential amino acid L-arginine, and when ingested supplants its role, creating canavanine proteins that don't function properly in the body. Eating seeds with L-canavanine led to L-arginine deficiency in McCandless that he wasn't able to reverse because of a lack of alternative dietary options as he became weaker and less able to hunt and forage.

It took 21 years of investigation by Krakauer to determine the identity of the toxic chemical in the seed that killed Christopher McCandless, illustrating the difficulty in determining the toxicology of unknown poisons. While the symptomology and chemical constituents of the most 'popular' plant toxins are relatively well known, there are many poorly known plants, especially those that form the basis of traditional medicine.[49,50]

* * * * *

We will finish this chapter by looking at some cases of deliberate amatoxin, mushroom poisonings.

Ancient Rome was not a healthy place for anyone—not even its rulers. In 54 AD, Emperor Claudius died from eating poisonous mushrooms at a banquet.[51] Over the centuries, there has been much speculation whether his wife, Agrippina, administered the poisonous mushrooms in a plot to get rid of him, as her son Nero from a previous marriage was coming of age. With Claudius's death, Britannicus, his own son by a previous wife, Messalina, and first in line, would not become Emperor.

Claudius's symptoms and sudden death are consistent with him eating fly agaric, *Amanita muscaria*, at the fatal banquet. Only Claudius died, and not the other guests because the levels of the psychoactive substance muscimol in this mushroom are generally not high enough to kill a healthy person. Fifteen mushroom caps would be the fatal dose for a healthy person. Claudius, however, suffered from dystonia, a neurological movement disorder, making him more susceptible than others to the effects of the toxin. By contrast, a single mushroom of the much more deadly death cap, *Amanita phalloides*, containing amatoxins and phallotoxins, would have been fatal, albeit over a period of several days.

An alternative explanation offered for Claudius's death is that the death cap mushroom poisoned him, but Agrippina felt that it was taking too long for him to die.[52] On instructions from Agrippina, his physician, Xenophon, induced vomiting by tickling the inside of his throat with a feather—a commonly known procedure used to help the sick purge their digestive system. The Roman historian Suetonius (*c.*69–after 122 CE) wrote in *The Twelve Caesars* that attendants would often tickle Claudius's throat to purge him after gluttonous banquets. But, in this case, Xenophon smeared the feather with a toxic overdose of an extract of the bitter gourd colocynth *Citrullus colocynthis*, which Roman physicians frequently used as a purgative. At high doses, bitter gourd can cause colitis, diarrhoea, and liver impairment.[53] Merchants had imported colocynth to Italy in Roman times from arid areas of the Middle East and so it was certainly available. The less exciting explanation for Claudius's sudden death is cerebrovascular disease. At the time of his death, he was 64—an old age for the time—and known to be in poor health.[54]

No matter the exact cause of Claudius's death, the case illustrates the extreme toxicity of some mushrooms. About 100 of the 100,000 known species of mushroom are reported to be toxic, although both numbers are increasing as fungal taxonomists identify new species. Reports of about 200–250 deaths per year worldwide are likely an underestimate, and these numbers may be increasing as poverty-stricken migrant populations travelling through Europe are forced to forage for food. Nowadays, a person can survive ingestion of toxic mushrooms if treatment is immediate. But in the past, this was not the case. And medical treatment is inadequate in some countries. Ironically, the plant toxin atropine from belladonna is an antidote in some cases of mushroom poisoning.

We need experts to identify mushrooms ingested in accidental poisonings. Unless you are an expert and absolutely sure, don't eat wild-collected mushrooms. There are too many toxic mushrooms that look just like edible ones.

Although wild-mushroom poisoning, or mycetism, is usually accidental, deliberate poisonings do occasionally occur. A review of 6,317 toxic mushroom ingestion cases over five years in California reported 731 (13%) intentional exposures.[55] In addition to Claudius, other famous mushroom homicides may include the deaths of Siddhartha Gautama (the Buddha) around 479 BCE, Pope Clement VII in 1534, and Holy Roman Emperor Charles VI on 20 October 1740. Nonetheless, each of these cases is disputed and may or may not have been deliberate.

One definite example of deliberate mushroom poisoning and homicide is the 1918 Paris *Affaire Girard* that took place under the shadow of the great spring offensive during the last year of the First World War. Originally trained as a druggist, the killer, Henri Girard, had abandoned that occupation to become an insurance agent. He took to murdering friends after insuring their lives, enlisting the help of his wife, his mistress Jeane Douétèau, a chauffeur named Bragnier, and a wine merchant, Rieu, as accomplices. In a murder spree, he injected some of his friends with the bacteria *Salmonella typhi* (typhoid), while the group fed others with dishes or drinks poisoned with wild mushrooms. A vagrant known only as le père Théo gathered the mushrooms from the forest of Rambouillet. Girard gave Théo instructions to gather amanitas with white gills, veil, and volva—characteristics of the deadly death cap. But it also shared these characteristics with the less deadly but similar-looking false death cap (A. citrina, also called A. mappa).

Depending upon which mushroom Théo had gathered, some guests went home and died, while others, fortunately, only had indigestion. One woman, Mme Monin, who died following mushroom poisoning, had four life insurance policies taken out on her, each worth 20,000 francs. On these policies, she named Girard or his accomplices as beneficiaries. A few days after Mme Monin died, Girard tried to cash in the policies, but one of the insurance companies became suspicious, leading to Girard's arrest. Girard died in prison awaiting trial, while his wife and mistress were both sentenced to long jail terms.[52,56]

* * * * *

The cases I have described here illustrate the common use of poisonous plants to commit murder. Other non-plant toxins, such as antimony and synthetic chemicals, are also used. Nowadays, the identity of at least the common poisons can be readily determined with chemical testing by pathologists. Yet this was not always the case, making poisoning perhaps more popular in the past than nowadays. The next chapter switches focus from the use of plants for nefarious crimes to crimes involving transporting rare or restricted plants, including illicit drug plants. Undertaking these crimes can be dangerous to those who take part, especially when the criminal underworld becomes involved.

HIDING IN PLAIN SIGHT

You can get off alcohol, drugs, women, food and cars, but once you're hooked on orchids you're finished. You never get off orchids . . . never.

Joe Kunisch, commercial orchid grower[1]

Michael Ormand (not his real name) had an obsession with collecting rare, carnivorous pitcher plants. His lawyer said that Ormand's obsession was a coping mechanism to help him manage symptoms of post-traumatic stress disorder, anxiety, and depression. The potted plants overrunning Ormand's small apartment were testament to his obsession.

But what started as a fascination with carnivorous plants had become an illegal obsession for the Portland State University student. What had begun with visits to local nurseries to buy a few plants turned into a social-media-fuelled, international smuggling operation. He joined an online carnivorous plant forum and a Facebook group for other collectors. When his Internet activities were no longer enough, he started hosting monthly meetings with other enthusiasts at a local pub in Portland, Oregon. His activities were open and at this point quite legal.

Ormand's illicit activities unravelled when a package addressed to him from a seller in Malaysia was intercepted at a mail-sorting facility in Los Angeles in October 2013. The package contained 35 wild-collected *Nepenthes rajah* plants with soil still clinging to their roots. Known as the giant Malaysian pitcher plant, *N. rajah* is the largest pitcher plant in the world, with pitchers or traps up to 41 cm in size (Plate 4). Some claim it is large enough to capture rats, although typically these carnivorous plants trap and consume insects. The carnivorous habit has evolved in several independent lineages of plants and represents a strategy to obtain nutrients in poor habitats. Carnivorous plants fascinated Charles Darwin; he famously referred to the Venus fly trap (*Dionaea muscipula*) as 'the most wonderful plant in the world'.[2]

Nepenthes rajah is extremely rare, occurring in the wild only on two mountain tops in Malaysian Borneo. It is internationally recognized as an endangered species, affording it the highest level of protection. Collecting it from the wild and shipping it across the world is a serious crime.

Police arrested Ormand after an undercover operation worthy of a drug sting by federal agents with the US Fish and Wildlife Service. They subsequently tried him for violations of the US Lacey Act, which makes it unlawful to import certain plants, including *N. rajah*, without permits. He had shipped $40,000 worth of plants into the United States between 2013 and 2015. However, despite the seriousness of this federal crime, Ormand was given only three years of probation, including six months of home detention. Ormand had expressed remorse at his trial, and the judge noted that his arrest and prosecution had 'got his attention'.[3, 4]

Collectors routinely uproot charismatic rare plants from the wild, smuggle them across international borders, and sell them to connoisseurs. You may be familiar with the term 'charismatic megafauna', used to refer to the wild animals that the public like and have an affection for (e.g. elephants, lions, blue whales, giant pandas). Likewise, plant collectors have their favourite, or charismatic, plants that get a lot of attention, including carnivorous pitcher plants, orchids, cacti, and cycads.

In addition, smugglers illegally harvest and ship around the world timber from rare and endangered trees. The illegal cutting of desirable timber species, such as rosewoods, hastens their extinction and worsens the ongoing ecological and environmental problem of deforestation. The wood from these trees is in high demand for luxury furniture. In 2004, the World Bank summarized the problem of illegal forest cutting succinctly: every two seconds, an area the size of a football pitch is illegally logged, 90% of logging is illegal in some countries, and this activity generates some $10–15 million annually that is circulating through criminal groups worldwide.[5]

Smugglers traffic illicit drug plants around the world from growers to users. Drug cartels are responsible for much of the shipping of drugs sourced from plants. These cartels focus on particular plants and geographic regions. For example, the Mexico based Sinaloa Cartel traffics heroin (from opium poppies) and cocaine (*Erythroxylum coca* and *E. novogranatense*) from across Central and South America into the United States and beyond. The former Medellin Cartel based in Colombia primarily trafficked cocaine and some marijuana from South America into Florida.

In this chapter, we will look at different kinds of plant trafficking across international borders, and how forensic scientists try to tackle this trade. Rare plants, invasive plants, and drugs are illegally poached and trafficked across borders, representing a major international problem. These activities are 'plant crimes', and are considered acts of 'biopiracy' when inadequate compensation is given to the country or Indigenous Peoples from which the plant is harvested.

* * * * * * * * *

Perhaps it's not usually considered a trafficking problem, but introducing invasive non-native plants that escape into new habitats disrupts and damages ecosystems, which threatens native species. The spread of these plants is a biosecurity issue—much like the spread of a virus during a pandemic—and everyone must treat it as such. Some of these invasive plants are so much of a problem that it is against the law for nurseries to sell them. In 2017 the US state of Ohio passed legislation listing 38 non-native, invasive plant species illegal to sell, offer for sale, propagate, distribute, import, or intentionally disseminate. Ohio's list included plants that had been popular ornamentals, including the wetland herb *Lythrum salicaria* (purple loosestrife) and the tree *Pyrus calleryana* (Callery pear). Such laws are widespread but difficult to enforce. [6,7]

A few years ago, one of my botanical colleagues found purple loosestrife for sale at a local nursery. She reported the nursery to the state Department of Natural Resources. It is easy to get these and other banned invasive plants online from multiple sellers. Between 2017 and 2019, investigators documented availability for purchase of 778 of 1,285 so-called regulated invasive plant species in the US from 1,330 vendors. [8]

A 'bad five' list of popular aquarium plants illustrates the problem associated with the transport of invasive, non-native plants across international borders. These plants are water hyacinth (*Eichhornia crassipes*), water lettuce (*Pistia stratiotes*), parrot's feather (*Myriophyllum aquaticum*), Kariba weed (*Salvinia molesta*), and red water fern (*Azolla filiculoides*)—all plants of South American origin—and they are all damaging freshwater systems. The native flora of South Africa is highly diverse and unique, with many endemic species that occur nowhere else on the globe. But the flora of South African freshwater habitats is threatened by the invasion of these non-native species, following their introduction by the local aquarium industry. The plants spread rapidly when introduced, often accidentally, into the wild. Sometimes this spread occurs when someone cleans out their aquarium and dumps the plants down the drain or directly into a stream, pond, or lake. These plants damage the integrity of native plant ecosystems, block waterways, and damage hydroelectric systems.[9]

Selling officially listed noxious weeds is a crime. The courts fined an Australian hardware retailer and four horticultural suppliers $15,000 in 2010 for selling Mexican feather grass (*Nassella tenuissima*, formerly named *Stipa tenuissima*) between January and May 2008. There have been other prosecutions for selling this ornamental grass. Australian law considers this grass a noxious weed, and it is illegal to import or sell it in the country. South Africa also lists it as a Category 1 weed, and California lists it as a noxious weed.

Nassella tenuissima is a popular ornamental grass because of its graceful, fine-textured inflorescences. It is a native plant from the south-western United States, Mexico, and Argentina, but is invasive in Australia. It readily escapes ornamental plantings and naturalizes in native pastures, displacing palatable and more nutritious native plants that cattle feed on. The seeds have awns (bristles) that contaminate wool produced from sheep grazing invaded pastures. Despite legislation banning its import and sale in Australia, it is still available on the Internet from international online traders. Vendors sometimes sell it under the old, outdated name

(*S. tenuissima*) or a misleading common name such as elegant spear grass, silky thread grass, or pony tail grass. Identification of this grass is important when biosecurity agents find it, and its recognition is facilitated by description of its key characteristics via social media postings and the development of DNA barcodes.[10,11,12,13]

* * * * *

The 1975 Convention on International Trade in Endangered Species of Wild Fauna and Flora (CITES), also known as the Washington Convention, protects rare plants to the extent that it can.[14] One hundred and eighty-three countries signed this convention aimed at protecting around 30,000 different types of plants (70% of which are orchids) and approximately 5,800 different animals. CITES assigns species different levels of protection depending upon their rarity.

'Appendix I' is the highest level of protection for species threatened with extinction. It prohibits any export or import for the 334 Appendix I plant species and four subspecies, except artificially propagated orchids. Appendix I lists seven species of orchid and two orchid genera.

'Appendix II' includes 29,644 plant species that, while not threatened with imminent extinction, require control in trading. These Appendix II plants include the entire Orchidaceae family of 870 genera and 25,000 species. Regulations for Appendix II listings only allow movement across international borders of species with an official permit.

Finally, 'Appendix III' species include those that have some level of protection in one country or another and require export permits from those countries. Mongolian oak (*Quercus mongolia*) for example, is an Appendix III listed species. A lumber company was caught illegally trafficking it from protected Russian forests instead of harvesting it legally from plantations in China before selling it on to a US hardwood flooring retailer. In this case, a court fined the lumber company, US hardwood flooring retailer Lumber Liquidators, Inc., $13.15 million in 2016 under the 2008 Lacey Act. As a part of the restitution required under the sentencing, Lumber Liquidators agreed to a $1.2 million community service payment to develop a wood-identification device that will help in future enforcement of illegal trafficking.[15]

The Mongolian oak case illustrates the need for CITES signatory countries to have their own laws to add teeth to the Convention and to allow enforcement and prosecution. Examples are laws such as the Canadian Wild Animal and Plant Protection and Regulation of International and Interprovincial Trade Act (1992), the European Union Timber Regulation (2010), the US Lacey Act (2008), and the Australian Illegal Logging Prohibition Act (2012). Each of these laws legislates protection within its respective countries. The countries from which smugglers take the plants also need their own legislation to allow domestic enforcement to stop the problem at its source. In Malawi, for example, where a lot of illegal logging occurs, legislators passed the Forestry Act Amendment Bill in February 2020. This act updates their decades-old Forestry Act, which was last updated in 1997. This amendment provides for increased conservation efforts, better regulation and law enforcement, and enhanced penalties and fines to help protect Malawi's forests.[16]

Stemming the tide of plant smuggling is difficult for many reasons. Listing a plant under CITES alerts authorities to the plants that they need to be tracking. But forensic testing of traded plant shipments is less than 1%. The identification rate is low, and there are errors in identification. Rare orchids not in flower are almost impossible to distinguish from common ones by looking at just leaves and tubers, as we saw in the Preface (Figure 1). And smugglers can deliberately mislabel the tubers of restricted plants as something legal. Lumber of a rare, CITES-listed tropical tree is difficult to spot when hidden in a pallet of legal lumber.

Once illegally sourced plants get into the legal commercial stream, the market opportunities expand enormously, with subsequent enforcement becoming even more difficult. Timber cut and milled into boards requires sophisticated microscopy, DNA barcoding, comparison databases, and machine learning computer technology to aid in identification. Enforcement is primarily at ports of entry, but it also depends upon control in source countries, countries along the way, and destination countries.

* * * * *

During woodworking class in school, I recall that the teacher gave me a piece of dark ebony (*Diospyros* spp.) to turn on a lathe to make the column for a lampstand. I remember the glossy, heavy feel of that piece of wood and how exotic it seemed to my young eyes. It was beautiful, and its musky smell was alluring. All I knew was that it had come from some tropical forest. I could not have known then that more than 100 species of *Diospyros* harvested for ebony would be CITES-listed and that the piece I had almost certainly came from a tree cut from a wild virgin forest. I also didn't realize that loggers had likely cleared the forest to make a sugar plantation, and the biodiversity it harboured was lost forever.

All that I didn't know as a young schoolboy is now well known, yet the plunder of rare, precious, and endangered timber from virgin forests continues. Rosewoods (*Dalbergia, Guibourtia,* and *Pterocarpus* spp., and a few other tropical trees in the Fabaceae legume plant family) are the most highly prized tropical timbers. *Dalbergia nigra* from Brazil is Appendix I-listed, and all the other 300 *Dalbergia* species were Appendix II-listed in 2016. CITES officials are listing other rosewoods regularly: they added *Pterocarpus tinctorius*, the African Mutluk from Malawi, in August 2019. However, many endangered rosewoods are not CITES-listed and thus have limited protection.

Most rosewood, sourced from African countries (including Madagascar) and Southeast Asia, is sent by poachers to China, with a smaller amount heading for Saudi Arabia. An undercover investigation in Nigeria and Madagascar carried out by the Environmental Investigation Agency revealed illegal harvesting of rosewood from National Parks and remote forests. Bribed, high-level government officials signed Nigerian shipping permits to China, blatantly listing the rosewood by name (*Pterocarpus erinaceus*—commonly known as 'kosso').

The sale of furniture fashioned from endangered trees is sometimes brazen. Would you pay $1 million for a four-poster bed? How about $80,000 for a shelving unit or $24,160 for a dining table with chairs? A high-end Chinese store openly advertised these prices in 2009 for Madagascar-sourced rosewood furniture. Collectors prize these endangered rosewoods for expensive furniture because of the rich, deep-red colouration of the wood. An undercover environmental activist posing as a customer wishing to ship rosewood furniture to the United States covertly recorded staff in a Chinese furniture store advising him to declare it as 'wooden furniture'. Showing their educated complicity, they told him not to mention *Dalbergia louvelii*, the Latin name of the violet rosewood from Madagascar. *Dalbergia louvelii* is a CITES Appendix II-, IUCN Red-listed endangered species.

The scale of these operations is staggering. From 2015 to 2017, smugglers exported more than 10,000 shipping containers with more than 1.4 million rosewood logs, which were valued at more than $300 million, from Nigeria to China in violation of CITES rules and Nigerian law. Bribed government officials who 'legalized' these shipments issued more than 4,000 permits, retrospectively.[17]

Worldwide, 35% of wildlife seizures from 2005 to 2014 were rosewood, comprising the single largest wildlife category—more than ivory from elephant tusks, which was second at 18%.[18]

The second most frequently seized rare timber plants after rosewoods are 'agarwoods' (*Aquilaria* and *Gyrinops* spp.). A dark, resinous substance is produced in the fragrant heartwood of these trees in response to infection by fungal pathogens (such as *Phialophora parasitica*), wind, lightning strikes, or other stresses. Manufacturers and artisans use the aromatic resin, called 'aloes', in perfumes, incense, and small carvings. While only 10% of agarwood trees produce the resin, illegal harvesting usually removes everything in its search, decimating native forests. The agarwood trees are CITES Appendix II-listed, Southeast Asian evergreen trees. Some of the agarwoods such as Walla Patta (*Gyrinops walla*) from western India and Sri Lanka have been provided additional protection at the national level as their commercial value has increased. [18, 19]

DNA barcoding and metabolomics are two of the forensic methods used to help protect agarwood. DNA barcoding reference databases (see Chapter 5) allow successful discrimination among several species in the *Aquilaria* genus. Metabolomic studies characterize the end-point or intermediate biochemicals (metabolites) produced by the plant because of enzymatic reactions. Analytical chemists can characterize a species by the presence and amount of these metabolites, which may include carbohydrates, amino acids, or organic acids. A metabolomics-based approach has allowed for the identification and authentication of agarwood wood chips of A. *crassna* and A. *malaccensis*. The latter is the main agarwood-producing species and was the first to be CITES listed, in 1995.

Ultimately, conservationists hope that use of a variety of forensic techniques will discourage illegal trade of agarwoods, while breeders develop domesticated varieties that foresters can plant in commercial plantations. [20, 21, 22]

Satellite sensing tied to smartphone apps used by community volunteers has shown promise in monitoring illegal deforestation. Trained monitors in the Peruvian Amazon forests received alerts on their smartphones when tree canopy loss was detected remotely, allowing them to address the issue locally through increased patrolling, or alerting the authorities. The programme led to a decrease in forest cover loss over two years. Beyond the value of harnessing advanced technology, this type of community volunteer programme empowers involvement of Indigenous Peoples in forest management and policy.[23, 24]

* * * * *

Michael Ormand and his obsession for rare pitcher plants are but one example of the problem of poaching and trafficking charismatic plants. Since Victorian times, collectors have wanted the most unusual, rare, and beautiful plants for themselves. In the 18th century, botanical gardens sent out collectors, such as Ernest Wilson, from Europe to the most remote parts of the world to collect plants. Wilson introduced an estimated 1,000 or more new plants to the Western world. Joseph Dalton Hooker, director of Kew Gardens, brought back some 7,000 species from the

Himalayas. European collectors prized orchids, with breeders hybridizing rare with common species.

Today, this orchid fever or 'orchidelirium' continues. While the trade is predominantly for cut flowers and greenhouse-grown plants, plants illegally collected from the wild are driving some species to extinction, fuelled by obsession, complicity among the parties involved, and fraud.

The discovery in 2009 of a beautiful orchid new to science was an occasion of great excitement. Local H'Mong (Meo) people had discovered an undescribed lady's slipper orchid growing on steep limestone cliffs in a restricted 0.35 km² area of primary forest in north-western Vietnam. Botanists named the orchid *Paphiopedilum canhii* in honour of a local man, Chu Xuan Canh, who could get specimens to flower in his garden. Dealers immediately rushed to make collections of the orchid for commercial trade, and the population of 10,000 to 15,000 mature and juvenile specimens is now 99.5% gone despite it being CITES-listed. There are plenty of other examples of orchid populations that dealers are decimating by over-collection. As one example, several lady's slipper orchids, such as the shiny green leaf Paphiopedilum (*Paphiopedilum glaucophyllum*) in Java, Indonesia, are absent now from most of the range. And *Phragmipedium kovachii* is virtually extirpated in its limited area of the Andean cloud forest of northern Peru, while the spectacular Grammangis orchid (*Grammangis spectabilis*) has only nine wild individuals left in its Madagascar habitat.[25, 26]

In 2001, orchid collector Matthew Kowal (not his real name) smuggled into the United States a live orchid purchased from a roadside vendor in the Amazon jungle of north-eastern Peru. It was clearly a lady's slipper orchid, but it was undescribed and new to science. After he took it to the Marie Selby Botanical Gardens, their taxonomists officially named it *Phragmipedium kovachii*, a new-to-science orchid. With its large, pink to purple flowers, it is a stunningly beautiful plant. But smuggling the plant into the United States implicated Kowal in a CITES treaty violation, as the orchid became listed as critically endangered by the International Union for the Conservation of Nature. At his trial, the judge fined Kowal and sentenced him to probation. And for their complicity, the judge revoked the Marie Selby Botanical Gardens' CITES permit to import listed plants.

The Kowal case and others like it illustrate the obsessiveness of some plant collectors, and their willingness to circumvent the law. Scientists are human, too, and they are not above greed and seeking personal gain. Eric Hansen raises the troubling assertions that botanical gardens may turn a blind eye to, or even encourage, illegal plant collecting to help bolster their collections in his controversial 2016 exposé *Orchid Fever: A Horticultural Tale of Love, Lust, and Lunacy*.[1, 27]

Obsession drives the collection of many rare plants. State agents caught orchid fanatic and nursery manager John Laroche and three Seminole Native Americans, Dennis Osceola, Vinson Osceola, and Russell Bowers, red-handed in 1993 with 92 orchids of some nine species. The stunningly beautiful ghost orchid, *Dendrophylax lindenii* (Plate 5) was one of the major foci of their collections along with some of the 45 other rare orchids that occur in the Fakahatchee swamp, a nature preserve 80 miles west of Fort Lauderdale, close to the Florida Everglades. The ghost orchid Laroche was collecting is endemic to five counties in south Florida and Cuba, and is CITES Appendix II-listed and state-endangered. The plant itself is a leafless perennial epiphyte (i.e. a non-parasitic plant growing on the surface of another plant) with a highly reduced stem and a mass of camouflaged photosynthetic roots—hence the name as the white flowers seem to hang in mid-air like ghosts.

Laroche and his Native American co-conspirators stuffed their plunder into pillowcases as they stepped out of the Fakahatchee swamp. The state charged them with violating a state code that forbids removal of plant life from state parks, specifically for cutting up trees and taking plants. The Seminole tribe had hired Laroche to guide them in finding rare orchids. Laroche believed that a loop-hole in the state code exempted native Seminoles from laws protecting rare plants. Nevertheless, Laroche was fined $500 and banned from the Fakahatchee Strand Preserve State Park for six months in an arguably light sentence that doesn't seem to fit the crime.[28] This detail is a small part of a much more involved story that attracted a lot of attention following the publication of Susan Orlean's 1998 book *The Orchid Thief* and the 2002 Hollywood movie *Adaptation*.[29]

One of the many conservation challenges in dealing with poaching and trafficking of orchids, and other charismatic plants including cycads and cacti, is identification of specimens that border-control agents examine at ports of entry. Smugglers will deliberately mislabel a CITES-listed rare species as a common, unregulated species. A second challenge is tracking the buyer–seller networks.

Geoffrey Tremont (not his real name) illegally imported *Phragmipedium* lady's slipper orchids to the United States in an obvious case of 'hiding in plain sight'. All 15–20 species of *Phragmipedium* orchids are CITES Appendix I-listed. Consequently, CITES regulations require permits to export and import any of these orchids across international borders. In this case, Tremont of

Texas collaborated with Marcelo Ruiz (not his real name) of Peru. Ruiz obtained permits of different species of artificially propagated orchids. He then substituted these with CITES Appendix I-listed *Phragmipedium* orchids, which Peruvian authorities did not allow for export. Tremont deliberately encouraged the mislabeling of these shipments with names of less strictly regulated orchids. Ruiz provided a key to Tremont of the true identity of the orchids he shipped. Tremont then knowingly falsified import documents and sold these CITES-protected orchids.

The indictment stated that in February 2003, Ruiz shipped approximately 1,145 orchids to Tremont, including 45 *Phragmipedium* and 445 specimens of Appendix II-listed orchids. In August the same year, he shipped 700 orchids, including an undisclosed number of *Phragmipedium* orchids and several Appendix II orchids. Customs agents valued these shipments at $86,947. Tremont was particularly keen to receive *Phragmipedium* orchids, and his true intent and complicity is revealed in letters he wrote to Ruiz. On 21 September 1999, he sent shipping instructions to Ruiz, writing, 'I don't see any problem with shipping Phrags *** Make sure they are wrapped with moss and paper and in plastic marked *Maxillarias* as before.'[30]

Maxillaria refers to the spider orchid genus, which would be Appendix II-listed with lesser export restrictions than that those of *Phragmipedium*. Tremont was fully aware of the CITES regulations that he was violating, as he stated in a letter to Ruiz on 30 July, 2000 (comments are verbatim):

> Please make sure that the plant has the appearance of being artificially grown (the styrofoam on the roots if possible). This would not be a good time to send any Phrags. As there are some regulations being changed regarding all Appendix I plants and they will be looking to try to find reasons to take plants.[31]

In other letters, he noted how customs agents usually only casually checked a few plants in a shipment and were often lacking expertise to confirm or dispute the names listed on the permits. Tremont and Ruiz had violated the US Endangered Species Act and the US Fish and Wildlife Service regulations. At trial, Tremont pleaded guilty, and the judge sentenced him to 17 months in prison, to be followed by two years of supervised release and a $700 fine. The sentence was affirmed following appeal. Ruiz was sentenced at his trial to 21 months, three years' supervised release, and a $5,000 fine.[31] Some commentators painted Tremont as a confused and innocent orchid collector, caught up by overzealous federal agents seeking to crack down on the orchid trade following the *Phragmipedium kovachii* case. However, the evidence presented at their trials suggests that both men knew exactly what they were doing. This case is a clear-cut example of how the illegal trade in rare plants can proceed.[30, 32]

*****

US federal agents tracked down the pitcher plant trafficking activities of Michael Ormand in part from an FBI investigation of his social media activities, which included an online forum and a Facebook group. Investigation of social media networks falls within the remit of digital (computer) forensics. This relatively recent addition to the forensic toolbox is based upon the recovery and investigation of data from electronic media that prosecutors can use in a court of law.

As with many things, it doesn't take too much time on the Internet to find illegal things for sale. Collectors, sellers, and buyers are using the Internet to trade in wild-collected rare plants. As evidence of this, a 2016 study of Internet-based social media networks among orchid-themed groups picked up 55,805 posts over a

12-week period, up to 46% concerning wild-collected orchids and 9% directly about plants for sale. It's clear that sellers are using online platforms to bypass CITES regulations. An anonymous online survey showed that 11% of 814 respondents had received orchids without required paperwork, with 10% of respondents admitting to having smuggled orchids. Lady's slipper orchids (*Paphiopedilum* and *Phragmipedium* spp.), including the critically endangered Venus slipper orchid (*Paphiopedilum kolopakingii*) from Borneo, were particularly popular in these illegal trades.[26, 33]

Similarly, an investigation of online nursery plant lists showed a clear willingness to stock and sell rare, wild-collected plants. Researchers found 65 nurseries from 14 countries to be taking part in e-commerce of 28 endemic plant taxa from the island of Crete, which is a biodiversity hotspot. Sixteen of the traded plants faced extinction risk, including the rare and beautiful red Cretan tulip (*Tulipa doerfleri*).[34]

The Russian *Q. mongolia* case described earlier in this chapter illustrates that underlying much of the illegal plant trafficking is fraud and misrepresentation. Three types of fraud in forest products occur: providing incorrect botanical identity of species (e.g. mislabelling rare species as something common); misrepresentation of geographic origin; and misrepresentation of product type (e.g. solid wood versus particle board). A survey found that 62% of tested wood products from US retailers were fraudulently labelled. But that was just part of the problem, as identifications were not helped by highly variable accuracy in forensic wood anatomy identification among 'experts'. Evidently, there's a need for more trained wood anatomists.[35]

Cacti are today's must-have plant for hipsters. Modern computer-based forensic technology is in a veritable arms race

with the poachers of these often rare plants. CITES agents investigated 11 persons in 2007 for geo-referencing via the Internet the
locations of individual rare cacti in Argentina, Peru, and Chile
before harvesting and then selling them on dedicated online
forums.[36] Worldwide, collectors often share GPS coordinates of
plant locations. To counter this activity, chief park ranger Ray
O'Neil inserted radio-frequency identification (RFID) microchips
into individual cacti including the iconic Saguaro (*Carnegiea gigan-
tea*) in the Saguaro National Park in Arizona. The hope is to
counter increasing thefts of these long-lived plants. As O'Neil
said in a 2009 interview, 'If you steal a cacti [sic], we will find
you.'[37]

* * * * *

Executives at Lumber Liquidators probably thought they were safe
in importing Mongolian oak from eastern Russian forests and
fraudulently labelling it as European oak (*Quercus petraea*). Who
would know the difference between these two oaks? Even expert
plant anatomists usually cannot tell apart different species of oak
using conventional microscopic examination. DNA barcoding is
needed to identify wood samples to the species level. Working out
the geographic origin of a sample is even more difficult.

Authorities caught out Lumber Liquidators by using the new
forensic technique of isotope ratio analysis. The German–British
company Agroisolab developed this technique based upon the
principle that trees absorb stable, non-radioactive isotopic forms
of carbon, hydrogen, and oxygen and rarer elements such as stron-
tium and sulphur into their wood. Isotopes of an element vary in
their number of neutrons, while the characteristic number of
protons that defines the element remains the same. Trees take
up differing amounts of isotopic forms depending upon the soil,

rainfall, altitude, and other environmental factors related to the location in which they grow. The ratios of the isotopes of these elements present in the wood are characteristic of the geographic region from which the plant takes up water. Researchers can measure these isotope ratios on wood samples using an isotope ratio mass spectrometer (IRMS). Isotope ratios in samples from annual rings can provide details going back in time. But to match a sample to a location requires a georeferenced database of known samples.

Agroisolab scientists enlisted help from conservationists to get Mongolian oak tree samples from 50 sites across Siberia to develop a reference database based on the analysis of isotope ratios. Comparison of the seized hardboard samples with those on the reference database showed that the wood came from Russian forests. Agroisolab scientists showed Lumber Liquidators' claim that the wood came from trees harvested in China to be false, and this evidence was key in the ultimately successful lawsuit against the timber company.[38, 39]

Scientists are developing other forensic wood techniques to help combat timber trafficking, including near-infrared spectroscopy (NIRS) and time-of-flight mass spectrometry (TOFMS). These techniques can help characterize unique chemical spectra possessed by the wood of different genera and species of trees. To appreciate this, imagine being able to chemically characterize the alluring aroma of fresh-cut pine.

Analysing the fibre of heavily processed pulp and paper wood products is another forensic technique in development. As we noted at the start of the book, wood has specialized fibre cells that vary at microscopic level among different tree genera and sometimes among species too. Experts can examine these fibres to tell if paper is sourced from illegal tropical hardwood. One study, for example, found that the wood used to make the paper in an office calendar came from Chinese forests, likely illegally.[40, 41]

In 1933, Syracuse University wood technologist Alexis J. Panshin noted that 'the woods of the *Swietenia* species cannot be separated anatomically with any degree of certainty'.[42] He was referring to the well-known problem that even highly trained botanists have difficulty telling apart tree species within genera based upon their wood anatomy. However, it is important to tell apart legally harvested species from protected species. The *Swietenia* studied by Panshin is the familiar mahogany. The wood of mahogany is light pink to dark-reddish brown with a satin or golden lustre. Woodworkers prize mahogany for its beauty and workability, and use it for fine furniture, musical instruments, and artisanal crafts.

Tereza Pastore and colleagues with the Brazilian Forest Service developed a portable NIRS scanner to tell apart Honduran mahogany (*S. macrophylla*) from other similar timbers, such as crabwood (*Carapa guianensis*), cedro (*Cedrela odorata*), cedrinho (*Erisma uncinatum*), and curupixá (*Micropholis melinoniana*).[43] But forensic scientists need additional methods to discriminate between the different species of mahogany in the *Swietenia* genus. Tuo He and colleagues from China, the US, and Brazil have recently developed computer-based machine-learning methods to tell apart two of the valuable, commercially important CITES-listed mahoganies, namely bigleaf mahogany (*S. macrophylla*) and American mahogany (*S. mahagoni*). These computer systems use inference-based algorithms that improve with experience to perform tasks without specific instructions.

In this case, researchers measured nine wood anatomy features microscopically from 263 specimens of *S. macrophylla* and

S. mahogoni. They could separate neither of the two mahogany species based on single features because of variability and overlap. But the supervised machine-learning algorithms were able to identify both *S. macrophylla* and *S. mahogoni* with a high level of accuracy, giving confidence in the use of this forensic approach.[44]

Less than 1% of globally traded timber is forensically tested. Nevertheless, forensic scientists hope that this will increase as methods become more widely available.[45] Commercial companies are now offering cutting-edge forensic testing services. These companies include Agroisolab,[46] which developed the stable isotope ratio analysis to characterize the geographic provenance of samples. In addition, Timtrace[47] in the Netherlands and Singapore-based Double Helix Tracking Technologies[48] are also offering forensic testing services. The World Forest ID project is working with these and other commercial companies, and London's Royal Botanic Gardens in Kew, to develop a tree DNA database that agencies can use in courts to prosecute criminal logging cases. Forensic taxonomists can then use DNA methods (see Chapter 5) to positively identify samples. In addition, they will store wood samples from across the globe at Kew to form part of Index Xylariorum, a global wood database. As project co-leader Phil Guillery said in an interview with the CNN TV network in 2019: 'With any wood product you will be able to test and know where it came from . . .. This is the first time where we have the ability of irrefutable evidence that can be . . . presented in a court.'[49]

In a landmark case, Double Helix Tracking Technologies (Double Helix) helped to prosecute the theft of bigleaf maple (*Acer macrophyllum*) from the Gifford Pinchot National Forest in the state of Washington in the United States. Bigleaf maple is a native deciduous tree in the Pacific Northwest of North America. The wood grain of bigleaf maple has distortions known as 'figuring' that make it prized in the musical instrument trade for making unique guitars. The US Forest Service realized that loggers had harvested more than $800,000 of figured bigleaf maple wood illegally and sold it to guitar companies across the United States. Double Helix worked with the US Forest Service, the University of Adelaide, and the World Resources Institute to build the first DNA profiling reference database for bigleaf maple.

The bigleaf maple database allowed investigators to match stumps in the forest to seized logs. To construct the database, Double Helix scientists conducted a genetic study using a next-generation sequencing (NGS) approach on 204 genetic markers, identifying 199 novel single nucleotide polymorphisms (SNPs) and five insertions/deletions (INDELS) from 65 individual trees. Ultimately, investigators at Double Helix sampled 400 trees and included them in the database.

At trial, defendant Henry Keating (not his real name), a 48-year-old wood buyer for J&L Tonewoods, and three others pleaded guilty to illegal interstate transport of wood products, specifically from bigleaf maple. This case was the first prosecution under the Lacey Act. US Forest Service personnel had advised Keating in 2012 that he needed to review permits when he purchased wood from loggers, but he ignored this advice. US attorney Annette L. Hayes noted that Keating was acting as a fence for stolen goods in buying illegally harvested wood in this case.

The judge sentenced Keating in 2016 to six months' jail time and a fine of $159,692. The three men who cut the wood and sold it to Keating also pleaded guilty to theft of government property. This case provides an example of the successful application of modern forensic methods to provide evidence to prosecute perpetrators involved in rare-species poaching and trafficking.[50,51,52]

The sustainable use of plants is all about the concept of 'traceability'. This concept relates to the ability to trace the history, application, and location of products, including the origin, processing history, and location after delivery. Traceability is essentially the chain of custody verification process that's well established and necessary in the handling of criminal evidence. Traceability is applicable here to illegally poached and trafficked plant materials. Originally developed for food safety concerns regarding CITES, traceability means having legal permits to document and record all stages in the commercial use of a rare plant. Ideally, land managers apply unique numbers or codes to plants *in situ*, to stumps, or to logs. The codes are physical markings, such as paint, waterproof tags, barcodes, or RFID tags. Conservationists and reserve managers add these records to a database, and customs and border-control agents can then check tagged plants at control points in the supply chain. The RFID tagging of Saguaro cacti in US deserts, as discussed earlier, is an example of this. Forest conservationists also put in place an RFID tag system for CITES Appendix II-listed ramin (*Gonystylus* spp.) timber monitoring in Malaysian peat-swamp forests between 2008 and 2010.

CITES requires submission of annual reports of all parties concerned with international trade of listed species. This documentation helps establish whether trade is within sustainable limits for a species. Of course, the system breaks down when smugglers obtain fraudulent permits, signed by bribed and corrupt officials, as in the rosewood and Mongolian oak cases, or when it is not checked, as in the bigleaf maple case.[53]

Corporations don't just smuggle rare trees for furniture. In 2017, the US Department of Justice fined Young Living Essential Oils, L.C. $500,000, plus $135,000 in restitution, and $125,000 in community service payments towards conservation, for knowingly harvesting and trafficking Brazilian rosewood oil (*Aniba rosaeodora*) and spikenard oil (*Nardostachys jatamansi*—a flowering herb) in Peru. Both plants are endangered and CITES listed, requiring permits which Young Living didn't get. The 1,900 litres of rosewood oil that the company imported to the US had a value of more than $9 million.[54] Many natural ingredients used in essential oils and cosmetics are listed under CITES, including orchids, cacti, certain aloe species, rosewood oil, agarwood oil, and black tree fern (*Cyathea medullaris*). To be fair, the cosmetics industry is aware of the issue, and offers primers and guidance to assist industry to try to ensure compliance with international regulations to promote sustainable harvesting.[55, 56]

* * * * *

The dealing in and shipping of illicit drug plants is the other side of the illegal plant-trafficking coin. There is a long list of illicit drug plants, including: cannabis (marijuana: *Cannabis sativa*); cocaine; doda, heroin, morphine, codeine, and other opiates from the opium poppy; kava (*Piper methysticum*); and kratom (a tropical evergreen tree, *Mitragyna speciosa*). These drugs are controlled substances under legal statutes. We will just look at cases related to three important plant-based illegal drugs—opium, cannabis, and cocaine—which give a flavour of the forensic issues and challenges that law enforcement faces in dealing with plant drugs.

Forensic identifications are made initially through presumptive or screening tests when these drugs are seized. Those identifications that can be presented in court are based on a confirmative test from chemical laboratory testing. Investigators also need to determine the origin of seized shipments, and they need to track them back to their source, either inside the country or across international borders.

Mislabeling is a ruse to fool customs agents when shipping illegal drugs, just as in the smuggling of endangered orchids and poached timber. In one recent case, the operator of an online botanicals business was convicted of importing and selling hundreds of misbranded kratom products into the United States. Federal Agents found that foreign suppliers had been instructed to mislabel kratom as 'incense', 'paint pigments', or other unregulated products. The kratom was then sold on the Internet as a non-prescriptive drug to combat various diseases, such as chronic pain, fibromyalgia, opioid withdrawal, and Lyme disease, and as substitutes for prescription drugs, without adequate advice to users on dosage and contraindications associated with the drug. While kratom is legal in many US states, it is not legal to make unsubstantiated medical claims about it without Food and Drug Administration clearance.[57, 58]

Farmers grow some drug plants in countries far from those in which they are eventually used—hence the trafficking by the drug cartels. The opium poppy plant is originally native to the eastern Mediterranean, although it is now naturalized worldwide. Producers use latex from the unripe seed pods to produce opium, which is a source of several opiates, including heroin and morphine. Cultivation for opium and its derivatives occurs largely in Afghanistan, Tasmania, Thailand, Laos, and Myanmar. By contrast, legal cultivation of opium poppies for poppy seeds is more widespread. Extraction of the opium drug from plants grown for poppy seeds is illegal.

'Poor man's heroin' or 'doda' is a highly addictive drug derived from the opium poppy. Although it is a dangerous drug, it is relatively easy to obtain and make. Producers grind the husks and unwashed seeds into a powder and dissolve them in water to make a tea-type infusion. Doda is a hazardous drug because it contains high levels of morphine, codeine, and thebaine, and its use can lead to severe reactions, and even death. Doda is especially popular in Canada, where users obtain it from flea markets and small South Asian grocery stores. A raid in three Canadian cities in 2016 led to the seizure of doda worth $2.5 million.

Prosecutions for possession of doda involve forensic laboratory testing, including morphological examination of plants, colour tests, thin-layer chromatography, high-performance liquid chromatography (HPLC), and gas chromatography. The spherical capsules topped by a stigmatic disk are relatively straightforward to identify as a poppy plant even in a dried state. Discrimination of the opium poppy with its lilac-white-pink petals from the non-opioid poppies with their red petals is easy when they are in flower. Other features, including clasping leaves on hairless stems, are also characteristic of opium poppies. However, powders and plant fragments are difficult to identify from visual characteristics. Law-enforcement officers can test suspected opium derivatives using a presumptive field colour test. One such tool is a one-step test in which an investigator places a small amount of sample into an ampoule containing Liebermann's reagent (potassium or sodium nitrate and sulphuric acid). This mixture elicits a wet-chemical reaction in which the colour indicates a particular opiate—yellow for heroin, red for morphine, orange for opium, and green for codeine.[59]

Similarly, marijuana (*Cannabis sativa*) is a plant illegal in many places but legal elsewhere. Many people often confuse marijuana with the related hemp, which is not a drug. While marijuana is now legal in many countries worldwide, it is still a Schedule I federally controlled substance in the United States, although it is legal to possess and use in several US states. The legal use of medical and recreational marijuana is widespread, albeit regulated, as is the use of closely related hemp and CBD (cannabinoid) varieties. Most scientists and regulatory agencies regard all three as being varieties of *Cannabis sativa* and not separate species of *Cannabis*, and growers increasingly plant and harvest them. The psychoactive constituent in cannabis is THC (tetrahydrocannabinol—an aromatic terpenoid), which occurs at effective concentrations (over 0.3%) mainly in the hairs on the flower buds of female *C. sativa* plants. Law enforcement needs accurate means to correctly identify whole marijuana plants, plant fragments (especially the THC-containing buds), and powders, and to discriminate them from hemp and CBD cannabis.

The distinction between marijuana ('pot') and hemp is important. Marijuana is more of a concern to law enforcement than hemp and CBD cannabis. A court jailed a Texas man for almost a month after police misidentified 3,000 lbs of hemp he was transporting in his pickup truck from California to New York City. The Texas arresting officer and local Drug Enforcement Agency officials were unaware of the difference between these two varieties of *C. sativa*. Farmers grow hemp plants for their fibre, and these are much taller than marijuana plants. Additionally, marijuana is defined as *C. sativa* with over 0.3% THC, whereas hemp and CBD plants have to have less. Under certain environmental conditions including drought and nutrient stress, hemp and CBD plants can have over 0.3% THC, in which case law enforcement considers them legally marijuana.[60, 61]

Law enforcement uses morphological and chemical presumptive tests to identify *C. sativa*. Morphological screening for *C. sativa* includes recognition of the characteristic palmate compound leaves with three to nine serrate leaflets, grooved stems, and flower buds on the female plants. Microscopic examination of the leaves seeks to find the characteristic long, curved cystolithic (i.e. calcium carbonate-containing) hairs on the upper surface and simple unbranched hairs on the lower surface. The glandular, THC-containing hairs occur on both leaf surfaces (and especially on the female flower buds). Chemical tests include microcrystalline colour tests and the Duquenois–Levine (DL) colour test. The DL colour test is a simple presumptive test for THCs; it produces a purple colour and is available in field kits for law-enforcement officers. Courts accept the results of the DL test as evidence.

Confirmative testing in forensic laboratories for cannabis is often a three-step process based on morphological examination of samples, microscopic examination for the leaf hairs, and a chemical analytical test using sophisticated instrumentation, such as HPLC or gas-chromatography mass spectrometry (GC-MS). Ultimately, analytical chemical testing of some sort is the *de facto* confirmative testing used for all illicit plants and synthetic drugs.

As we have seen, forensic investigators use DNA testing and the development of DNA databases in investigations in which identification of the source or variety of a drug is necessary. Indeed, you may recall from Chapter 5 that DNA testing for cannabis was the first botanical DNA evidence accepted in UK courts.

As we saw with the case of Geoffrey Tremont's illegal orchid smuggling, hiding in plain sight is the tactic often used to smuggle illegal plants across borders, and this is true for drugs such as cocaine. This highly addictive stimulant is made from the leaves of the coca plant (*Erythroxylum coca* and *E. novogranatense*) and transported as dried bricks in its processed form. Producers make these bricks by first soaking the harvested leaves in gasoline. They then drain away the gasoline, after which they dissolve the dried base in a solvent (often battery acid—weak sulphuric acid) before drying it again into crystallized bricks. The drug cartels then smuggle these bricks from the source countries in South America (especially Colombia, as well as Bolivia, Chile, and Peru) and into the United States for global distribution. Smugglers often bring the cocaine into Florida by boat, by truck at border crossings from Mexico into the south-western states such as Texas or New Mexico, or via a 'mule' by air into international airports. There are many imaginative ways of hiding the bricks, including in secret compartments in vehicles and luggage, in shampoo bottles and coolers, and under the clothes or in the body cavities of smugglers. In one audacious case, an attempt was made to smuggle $30,000 of cocaine in the shafts of hollowed-out golf clubs. Mules have died after drug capsules that they had swallowed burst inside their bodies.

One of the largest busts was the seizure of 39,525 pounds of cocaine with a street value of $1.3 billion in 2019. US Customs and Border Protection agents discovered the cocaine in shipping containers on board the *Gayane*, a 1,030-foot cargo ship owned by MSC (Mediterranean Shipping Co.), a Swiss shipping company flying under a Liberian flag, while it was at a stopover in the US port of Philadelphia. If laid end to end, the seized bricks of cocaine would have stretched 2.5 miles.

Agents discovered the cocaine following a routine inspection of the ship. Seals on seven of the steel shipping containers on board didn't look right. Upon further investigation, they were revealed to contain cocaine. The smugglers brought the cocaine on board the ship from 10 small boats that approached the ship at night while it was at sea in the Pacific Ocean, sailing from Chile and Peru towards the Panama Canal. They stuffed bales of cocaine bricks into containers they had illegally opened, and then put replacement seals onto the containers. After the stop in Philadelphia, the *Gayane* was due to sail across the Atlantic Ocean to ports in Europe, where dealers would have distributed the cocaine.[62]

We don't know which particular presumptive tests the agents used to identify the cocaine on board the *Gayane*, but investigators reported that the white powdery substance they examined tested positive for cocaine. Nevertheless, agents use sniffer dogs or handheld vacuum devices to detect the scent of the cocaine. Presumptive field tests are available for border-control agents, including colour and microcrystalline tests. If law enforcement seizes unprocessed coca leaves, then they can identify them because of their broadly elliptic shape similar to bay leaves (*Laurus nobilis*). The undersides of coca leaves are characteristically grey-green with two lines parallel to the midrib. Confirmative chemical lab testing of processed cocaine uses a variety of methods including thin-layer chromatography and GC-MS.[63]

* * * * *

In this chapter, we have looked at some of the main plants that criminals transport illegally across the globe. Smugglers transport these plants because they are desirable, rare ornamentals or timber, or non-native but invasive garden plants, or illegal drugs. In most cases, smugglers process these plants, making identification extremely challenging, or they transport them in a vegetative form, making their distinction from similar but legal plants next to impossible. These cases require sophisticated but time-consuming and expensive microscopic, chemical, or molecular tests requiring a high level of technology and expertise. Traffickers are well aware of these challenges faced by border-control and customs agents, and they play an odds game that is in their favour.

CODA: CONQUERING PLANT BLINDNESS

> As a rule, the more bizarre a thing is the less mysterious it proves
> to be. It is your commonplace, featureless crimes which are really
> puzzling.
>
> Arthur Conan Doyle, *The Adventures of Sherlock Holmes*[1]

We've seen in this book how plants can provide sometimes critical although seemingly cryptic evidence to help solve crimes. Indeed, plants can be very important in many aspects of the legal system. They can be used in the very act of committing a crime (e.g. poisonings), their presence can link suspects to crime scenes (Locard's *'les poussieres'*), and they can be processed to produce illegal and illicit substances (e.g. plant-based drugs) that can get individuals into trouble with the law. Moreover, many highly sought-after and valuable plants are illegally removed from their natural habitats, transported, and sold. In all of these cases, identification of the plants, plant fragments, pollen, diatoms, spores, or plant extracts requires specialist knowledge. This is where botanical forensics gets involved.

The solving of the most heinous of crimes committed by serial killers can be aided through the inclusion of botanical evidence. As Edmond Locard recognized at the turn of the 20th century, no matter how notorious the criminal or how carefully they plan their crime, they leave microscopic traces in the form of pollen and spores that can become critical evidence.

Yet, in spite of the examples described in this book, the inclusion of botanical evidence in criminal and civil trials remains relatively rare. Some argue that this is due to the lack of botanically trained CSIs. Obvious plant fragments may catch the eye of the CSI as in the *Sphagnum* peat moss murder case, or may prompt the need for a DNA investigation as in the seed pods found in the suspect's truck in the Maricopa case. But the botanical evidence and requirement for expertise needs to be spotted by the CSI in the first place as soon as a crime scene is established, before potential evidence is destroyed or lost.

In the Maricopa case, it was the earlier development of human DNA forensic testing that prompted investigators to ask whether plant DNA could be used similarly. Botanical forensics benefits from advances in other areas, but can itself pave the way. Improved chemical testing of plant poisons is an example. New developments are allowing many aspects of forensic botany to move forward. Improved chemical testing, provenance testing, DNA barcoding and profiling, development of plant DNA databases, and development of eDNA profiling are areas of rapid growth allowing more sophisticated forensic tracking and tracing of materials. These advances offer more opportunities for botanical evidence to be of value in the legal system.

In addition to the need for botanical expertise at crime scenes, more botanical knowledge is needed at ports of entry to identify illegal drugs, poached plants, and plant products, such as furniture made from endangered rainforest trees. Presumptive tests allow rapid on-the-spot tentative identification of plant drugs such as cocaine, but as we have seen, identification of poached orchids transported in the vegetative state without flowers is extremely difficult and time consuming. Processing samples for microscopic anatomical or DNA analyses (when DNA databases exist) is time consuming and expensive, so border agents need training to determine which plants might need such advanced testing.

Many universities offer a high level of training for botanists, but botany programmes are declining in academia, and the remaining programmes have diversified their offerings as molecular methods become more important, at the expense of in-depth 'classical' training in plant diversity, morphology, anatomy, and systematics. As an undergraduate botany student in the 1970s I took separate, advanced courses in vascular plant systematics, ferns, bryophytes, algae, fungi, palynology, morphology, and anatomy. Even then, I often felt underprepared when my mother, a keen amateur wildflower enthusiast, asked me the names of local wildflowers. Today, the botany department where I studied is no longer, while, as I remarked earlier, at my current institution the Department of Plant Biology has been absorbed into a School of Biological Sciences. The course

I offer on Forensic Botany is unique, but we no longer offer courses on algae, bryophytes, or palaeobotany.

Still, all is not doom and gloom for botany. Dedicated students can get first-class training; they just have to look around to find the best programme. Rather, when it comes to forensic science, what is needed is for the legal profession at all levels to embrace the value of botanical forensics and demand the hiring of trained botanists into forensic laboratories to work alongside the chemists, DNA experts, ballistics experts, and tyre-tread and fingerprint specialists. CSIs need to be trained to recognize the potential for botanical evidence to be important as they work on crime scenes. I am convinced that if more demands for the investigation of potential botanical evidence were made ('What pollen is in the mud of the suspect's shoe?'), then there would be more training of forensic botanists. Such a demand will drive supply.

The enduring popularity of gardening and natural history is testament to the love that many members of the general public have of plants. Natural history and native plant clubs and societies abound with enthusiastic and knowledgeable members. And yet, paradoxically, plants are often over-looked, taken for granted, and are too much part of the background. This 'inability to recognize the importance of plants . . . in human affairs' is referred to as 'plant blindness'.[2] Conquering plant blindness is not only important for conservation of the natural world, but necessary for better and more informed forensic investigations. For example, there's a need to broaden the scope of illegal wildlife trade policies to consider a greater variety of plants beyond the emphasis on timber trees (see Chapter 7), and to highlight plants as wildlife.[3]

Consider the words of fictional attorney Jackie Flowers, planning her defence of a suspect in Stephanie Kane's novel *Seeds of Doubt*: 'I've got a few surprises up my sleeve. . . . A forensic botanist to blow his autopsy report out of the water.'[4]

GLOSSARY

This glossary provides a brief explanation of some of the forensic chemical and molecular methods mentioned in the text. Identification of more compounds and improved genetic resolution of individuals and species from smaller and more degraded forensic samples are constantly improving through advances in methodology.

AFLPs (amplified fragment length polymorphisms)—A molecular method used for DNA profiling. Restriction enzymes are used to digest the DNA, and a subset of the fragments are amplified using specific primers. Presence or absence of the amplified fragments are compared using electrophoresis to help characterize individuals.

Chromatography—A method allowing the relative proportions of chemical components to be determined through separation of mixtures of chemicals (the analyte) as they pass as a solution (liquid chromatography) or in a gas phase (gas chromatography) through a liquid or solid phase (the 'column'). The constituents move at different rates based upon retention in the column leading to their separation. A detector records the concentration of analytes and a solvent in the mobile phase as they exit the system, providing a plot of analyte signal versus retention time. Chromatography is often coupled with MS to provide chemical separation before chemical characterization.

CRISPR (clustered regularly interspaced short palindromic repeats)—A powerful gene-editing technique in which an RNA molecule guides an endonuclease enzyme to cut, or cleave, DNA at specific locations. These locations are clustered repeated DNA sequences interspaced in bacterial genomes derived from bacteriophages (viruses) that had previously infected the organism. Scientists can adapt this phenomenon to edit target genomes.

DNA barcoding—Provides identification of plant or fungal species based upon variation within the *mat*K and *rbc*L plastid genes, *trn*H-*psb*A nuclear spacer plastid genes regions, and/or the nuclear ribosomal ITS gene regions.

DNA metabarcoding—Similar to DNA barcoding but allows the simultaneous identification of many organisms in an eDNA sample through next-generation sequencing.

DNA profiling (DNA fingerprinting)—Provides identification and tracking of individuals within a species based upon variation among DNA fragments. Historical prominence of methods flowed more or less in this order: RAPDs, AFLPs, microsatellites, and SNPs.

eDNA (environmental DNA)—DNA extracted from environmental samples such as soil or stream, lake, or ocean water. The DNA barcoding is used to determine the identity of the organisms that it came from and are assumed to be living associated with the sample.

Fourier transform infrared (FTIR) spectroscopy—Used to identify organic materials. Absorption of infrared (IR) radiation is measured using a spectroscope over a wide range of wavelengths. A Fourier transform mathematical process is used to convert the raw data to absorption spectra. The resulting spectrum is essentially a 'molecular fingerprint' of the molecule that can be compared to known reference samples.

Gas chromatography (GC or gas–liquid chromatography, GLC)—A form of chromatography in which the mobile phase of the mixture is carried through the column in a gaseous phase.

Gas chromatography–mass spectroscopy/spectrometry (GC–MS)—A tandem coupling of GC to separate chemicals in a mixture and then identify them using MS.

High-performance liquid chromatography (HPLC)—Liquid chromatography utilizing high pressure to force the mobile phase through the column.

Insertion/deletions (INDELS)—Insertions or deletions of nucleotide bases in the DNA of an organism. Indel variants of multiples of three nucleotides can shift the reading frame so that the DNA sequence codes can add or subtract sets of amino acids, or if not in multiples of three, prematurely stop the coding of protein synthesis. Derived from a single mutation event, indels are useful genetic markers for forensic identification.

Internal transcribed spacer (ITS)—A non-functional spacer gene region located between structural ribosomal RNAs (rRNA) useful for DNA barcoding. There are two ITSs in plants. The fungal ITS region is particularly variable and used as a universal fungal barcode sequence.

Isotope ratio analysis (isotope-ratio mass spectrometry, IRMS)—A technique that uses an isotope-ratio mass spectrometer (MS) to measure relative abundance of stable isotopes of an element (e.g. of C, N, O, H, S, Sr, and Pb) in a sample. IRMS is useful in forensics to track drug trafficking routes as the isotope ratio can be characteristic of where a plant grew.

Liquid chromatography (LC)—Chromatography in which the mobile phase is a liquid. See chromatography, HPLC, and LC/MS.

Liquid chromatography tandem mass spectrometry (LC/MS)—A tandem coupling of LC to separate chemicals in a mixture and then identify them using mass spectroscopy.

Machine learning—The application of artificial intelligence (AI) computer algorithms that access data and improve through experience. Forensic applications include programs that improve with time in their ability to discriminate among images of different timber species.

Mass spectroscopy/spectrometry (MS)—An analytical technique in which a sample is transferred into an ionizing chamber where it is fragmented by bombardment with electrons into ions, separated based upon their mass-to-charge ratio, allowing their chemical and molecular structure to be determined through reference to the signal of known structures.

matK—The maturase K plastid protein-coding gene useful in DNA barcoding of plants.

Metabolomics—Identification of intermediate biochemicals (metabolites) produced through enzymatic reactions. Particular metabolites can be specific for a particular plant.

Microsatellites (SSRs (simple sequence repeats), STRs (short tandem repeats))—Short stretches of DNA that consist of a sequence of bases that is repeated like an accordion, over and over again. The number of repeated bases can be used to generate a DNA fingerprint characteristic of an individual.

Near-infrared spectroscopy (NIRS)—This method characterizes chemical absorption spectra when exposed to near-infrared electromagnetic energy. NIRS has forensic value as it can be used to discriminate between genera, species within genera, and individuals of the same species from different geographic provenances.

Next-generation sequencing (NGS, also called massively parallel sequencing: MPS)—High-throughput, rapid sequencing of thousand to millions of DNA molecules simultaneously. Various competing technologies include Illumina™, PacBio™, and Oxford Nanopore™. NGS allows rapid sequencing of genomes at a low cost. NGS is valuable forensically for building DNA databases to identify and compare individuals within a species, for example, to determine the provenance of illegally logged timber.

Nucleotides—The organic building blocks of RNA and DNA consisting of a 5-carbon sugar molecule (ribose in RNA or deoxyribose in DNA), a phosphate, and a nitrogen-containing base. Nucleotides are linked together into long chains to form RNA and DNA.

Polariscope—An instrument that measures the angle of rotation of polarized light passed through an optically active substance. The polarized light will be rotated to the right or the left. This instrument helps determine the purity of samples based upon rotation of the polarized light which can be specific for a chemical. It can also be used to identify unknowns by reference to optical properties of known chemicals. Commonly used in the precious gem, food, beverage, and pharmaceutical industries.

Polymerase chain reaction (PCR)—Developed in 1985, PCR allows the 'bulking up' of the small amount of DNA in cells to provide a large enough sample for analysis. Alternate heating and cooling cycles cause the two complementary strands of the DNA molecule to separate at high temperatures, allowing DNA polymerase enzymes to synthesize a copy of each strand at the lower temperature. Repeated heating and cooling cycles double the amount of DNA each cycle, exponentially increasing the amount of DNA over multiple cycles.

Radio-frequency identification (RIFD) tags—Microchips that can be inserted into plants or animals to allow tracing. RIFD tags are forensically used to tag and track rare plants such as cacti.

Randomly amplified polymorphic DNA (RAPD)—Random fragment of DNA produced through PCR amplification with arbitrary short primers. Similarity among individuals of a species is determined by comparison of these fragments. The first forensic application of plant DNA profiling used RAPDs.

rbcL—Ribulose-bisphosphate carboxylase oxygenase chloroplast gene region used for DNA barcoding. This gene codes for the large subunit of RuBisCo, the enzyme that catalyses the uptake of CO_2 in most plant photosynthesis.

Sanger sequencing—A method developed by Nobel laureate Frederick Sanger in 1997 for determining the nucleotide sequence of a DNA sample. Although largely superseded nowadays by NGS, Sanger sequencing remains useful for short DNA sequences and NGS validation tests.

Single nucleotide polymorphisms (SNPs)—Single nucleotide differences ('molecular typos') occurring in stretches of DNA that arise during cell division as the DNA is copied. For example, the nucleotide base cytosine may be replaced with the base thymine in a particular portion of the DNA of an individual. SNPs can be unique to individuals and are used in DNA profiling of plants and animals.

Stable isotope analysis—Stable isotopes are naturally occurring alternative forms of elements (especially of C, H, O, N, S, and Sr) that are incorporated into phytochemicals as they are synthesized. These isotopes do not decay radioactively so their amount does not decrease through time in an organism. The amount of a particular isotope is influenced by the environment, climate, and geology, enabling the isotopic signature for the species in an area to be determined. Used in forensics to track the geographic provenance of a plant such as illegally harvested timber.

Thin-layer chromatography (TLC)—Chromatography in which non-volatile mixtures are separated based upon their electrical charge and travel across a sheet of glass, plastic, or aluminium foil coated with silica gel, aluminium oxide, or cellulose. This technique is especially important for separating and identifying multi-component formulations in the food and cosmetics industries.

Time of flight mass spectroscopy (TOFMS)—A method of MS in which the ion's mass-to-charge ratio is determined by measurement of the time taken to travel a fixed distance in a vacuum chamber.

ENDNOTES

Preface

1. Darwin, C., *On The Origin of Species*. 1910 reprint of 1st edition 1859, London, Melbourne, & Toronto: Ward Lock & Co Limited.
2. Corner, E. J. H., *The Life of Plants*. 1964, Chicago, USA: University of Chicago Press. p. 376.
3. Mabey, R., *The Cabaret of Plants: Botany and the Imagination*. 2015, London: Profile Books. p. 352.
4. Silvertown, J., *Demons in Eden: The Paradox of Plant Diversity*. 2005, Chicago, USA: University of Chicago Press. p. 192.
5. Stafford, F., *The Long, Long Life of Trees*. 2017, New Haven, Connecticut, USA: Yale University Press. p. 296.
6. Gibson, D. J., Obituary: Eric Vernon Watson BSc, PhD (1914–1999). *The Bryologist*, 2001. **104**(3): pp. 471–2.
7. Wandersee, J. H. and E. E. Schussler, Preventing plant blindness. *American Biology Teacher*, 1999. **61**(2): pp. 82, 84, 86.
8. Shakespeare, W., *Romeo and Juliet*. 1591, New York: Nelson Doubleday, Inc., Garden City.
9. Hill, M. O., Peter Greig-Smith (1922–2003). *Bulletin of the British Ecological Society*, 2003. **34**(4): pp. 10–11.
10. Clapham, A. R., T. G. Tutin, and E. F. Warburg, *Flora of the British Isles*, 2nd edition. 1962, Cambridge: Cambridge University Press. p. 1269.

Chapter 1

1. Hooke, R., *Micrographia: or Some Physiological Descriptions of Minute Bodies Made by Magnifying Glasses. With Observations and Inquiries Thereupon.* 1665, London: The Royal Society.
2. Graham, S. A., Anatomy of the Lindbergh kidnapping. *Journal of Forensic Sciences*, 1997. **42**: pp. 368–77.

3. Hall, D. W. and W. Stern, Plant anatomy, in *Forensic Botany: A Practical Guide*, D. W. Hall and J. H. Byrd, editors. 2012, John Wiley & Sons Ltd: Chichester, West Sussex, UK. pp. 119–26.

4. Lindbergh, A. M., *Hour of Gold, Hour of Lead: Diaries and Letters of Anne Morrow Lindbergh, 1929–1932*. 1973, New York: Harcourt Brace Jovanovich.

5. Hertog, S., *Anne Morrow Lindbergh: Her Life*. 1999, New York: Nan A. Talese, Doubleday.

6. Gribben, M. *The Yule Bomber*. The Malefactor's Register. Undated [cited 2020 2/4/2020]; available from: https://malefactorsregister.com/wp/the-yule-bomber–2/.

7. Ross, A. T. *CSI Madison Wisconsin: Wooden Witness. Peeling Back the Bark*. Koehler, Arthur. 1935 Radio Interview: https://fhsarchives.wordpress.com/2009/03/31/csi-madison-wisconsin-wooden-witness/ (accessed 12 August 2018). 2009.

8. Esau, K., *Plant Anatomy*, 2nd edition. 1965, New York: John Wiley & Sons.

9. Gaby, L. I., *The Southern Pines*. 1985, United States Department of Agriculture, US Forest Service. FS-256. https://www.fpl.fs.fed.us/documnts/usda/amwood/256spine.pdf (accessed 12 August 2018).

10. Lindbergh, A. M., *Locked Rooms and Open Doors: Diaries and Letters of Anne Morrow Lindbergh, 1933–1935*. 1974, New York: Harcourt Brace Jovanovich.

11. 'States Rush Action On Kidnapping Laws', in *New York Times*. 1932.

12. The Federal Kidnapping Act of 1932, 18 U.S.C. § 1201. 1932.

13. Delmont, R. *Lindbergh Kidnapping Hoax*: http://www.lindberghkidnappinghoax.com/ (accessed 21 January 2020). 1998–2017.

14. Scaduto, A., *Scapegoat. The Lonesome Death of Bruno Richard Hauptmann*. 1976, New York: G.P. Putnam's Sons.

15. Klein, L. (producer and director), *Who Killed Lindbergh's Baby?* 2013, PBS-NOVA, Season 40, Episode 4. A NOVA production by Lawrence Klein Productions, LLC for WGBH.

16. Sachs, J. S., *Corpse*. 2001, Cambridge, MA: Perseus Publishing. p. 270.

17. Rule, A., *The Stranger Beside Me*. 1980, W. W. Norton and Company. p. 480.

Chapter 2

1. Locard, E., *Manual de Technique Policière*. 1923, Paris: Payot.

2. Nordheimer, J., All-American Boy on Trial, in *The New York Times*. 1978.

3. *Bundy v State*, 471 So. 2d 9, 23 (Fla. 1985). Appeal to Florida Supreme Court for Leach conviction.

4. Burstein, J., CSI with a botany degree: plants can help solve crimes, in *South Florida Sun Sentinel*, 31 January 2011.

5. *Frye v. United States*, 293 F. 1013 (D.C. Cir. 1923), 3968 1923.

6. *Daubert v. Merrell Dow Pharms., Inc.*, 509 U.S. 579. 1993.

7. Mazévet, M., *Edmond Locard: le Sherlock Holmes français*. 2006, Brignais: Editions des Traboules.

8. Heinrich, E. O., Review [Untitled]. *Journal of Criminal Law and Criminology* 1932. **22**(6): pp. 939–40.

9. Locard, E., *Traité de Criminalistique: Vol 2 Les Empreintes et Les Traces dans L'enquete Criminelle*. 1931, Lyons, France: Joannés Desvigne et ses Fils.

10. Wagner, E. J., The French Connection of Sherlock Holmes, in *EJDissecting Room*. 2011: https://ejdissectingroom.wordpress.com/2011/02/25/the-french-connection-of-sherlock-holmes/ (accessed 8 January 2020).

11. Jackson, R. L., *Criminal Investigation: A Practical Textbook for Magistrates, Police Officers and Lawyers*, 5th edition. 1962, London: Sweet & Maxwell Limited. p. 448.

12. *Davis v Ducart*, No. 2. 13-cv-02570-JKS (E.D. Cal. Jan. 2, 2015).

13. The National Court Rules Committee. *Federal Rules of Evidence. 2020 Edition, Article VII—Opinions and Expert Testimony*. Pub. L. 93–595. 2015–2020; Available from: https://www.rulesofevidence.org/article-vii/ (accessed 18 January 2020).

14. *Buckley v Rice-Thomas*, Plowden 118, Court of Common Bench, 6 ConLR 117. 1554.

15. Milroy, C. M., A brief history of the expert witness. *Academic Forensic Pathology*, 2017. 7(4): pp. 516–26.

16. *Folkes vs Chadd*, 3 Doug. 157, 99 Eng. Rep. 589. 1782.

17. Wiltshire, P., *The Nature of Life and Death: Everybody Leaves a Trace*. 2019, New York: G.P. Putnam's Sons.

18. *Marlow v Douglas County*, No. 31013–2 (Wash. Ct. App. 2013) Unpublished, unsuccessful appeal. 2013.

19. Anonymous, Casey Anthony murder trial defense calls forensic botanist to testify after delay. https://www.washingtonpost.com/national/casey-anthony-murder-trial-defense-calls-forensic-botanist-to-testify-after-delay/2011/06/21/AGBtZaeH_story.html (accessed 18 January 2020). in *Washington Post*. 2011.

20. Colarossi, A., Casey Anthony's defense team attacks reliability of two state witnesses, in *Orlando Sentinel*, 29 March 2011. https://www.orlandosentinel.com/news/os-xpm-2011-03-29-os-casey-anthony-trial-new-motions-20110329-story.html (accessed 18 January 2020). 2011.

21. *State of Florida v Casey Marie Anthony*, Case No. 48-2008-CF-15606-O (Circuit Court of the Ninth Judicial Circuit in and for Orange County, Florida). 2011.

22. Florida Exotic Pest Plant Council. *List of Invasive Plant Species*. https://www.invasive.org/species/list.cfm?id=74 (accessed 18 August 2021). 2019.

23. Crandall-Stotler, B., Personal communication with the author. 2019.

24. *People v Ashley*, 566 N.E.2d 745 (1991).

25. Frye in the Trunk of a Car (Anthony case) https://lawprofessors.typepad.com/evidenceprof/2011/06/frye-in-the-trunk-of-a-car-antho ny-casc.html (accessed 8 August 2020), in EvidenceProfBlog, C. Miller, Editor. 2011.

26. Weiss, K. J., C. Watson, and Y. Xuan, Frye's backstory: a tale of murder, a retracted confession, and scientific hubris. *Journal of the American Academy of Psychiatry and the Law* (online), 2014. **42**(2): p. 226.

27. The National Court Rules Committee. *Federal Rules of Evidence. 2020 Edition, Article VII—Opinions and Expert Testimony*. Pub. L. *93–595*. 2015–2020; available from: https://www.rulesofevidence.org/article-vii/ (accessed 18 January 2020).

28. Bendectin. https://www.bendectin.com/en/ (accessed 16 January 2020).

29. *Mekdeci v Merrell Nat'l Laboratories, Div. of Richardson-Merrell, Inc.*, 711 F.2d 1510. 1983.

30. *Daubert v. Merrell Dow Pharms., Inc.*, 509 U.S. 579. 1993.

31. *Daubert v Merrell Dow Pharms.,* 951 F.2d 1128. 1991.

32. Raum, B. A., Expert Evidence, in *Forensic Botany: A Practical Guide*, D. W. Hall and J. H. Byrd, editors. 2012, John Wiley & Sons, Ltd: Chichester, UK. pp. 79–92.

33. *Flanagan v State of Florida*, 625 So. 2d 827 n.2 (Fla. 1993), p. 828. Establishes the 'pure opinion rule'. 1993.

34. *The Queen v Bonython*, 38 SASR 45. Establishes a two-part test based upon *Frye*. 1984.

35. *Harris v. Cropmate Co.*, No, 4-98-0269 4th Dist. 1/26/99. Crop herbicide case citing *Frye* and *Daubert*. 1999.

Chapter 3

1. BBC. Plant detectives: how bramble and co. can help solve crimes. https://www.bbc.co.uk/programmes/articles/5q2xGXDZvoS7hg3KQ l11vNg/plant-detectives-how-bramble-and-co-can-help-solve-crimes (accessed 27 January 2020). 2020.

2. Mazévet, M., *Edmond Locard: le Sherlock Holmes français*. 2006, Brignais: Editions des Traboules.

3. Locard, E., *Traité de Criminalistique: Vol 2 Les Empreintes et Les Traces dans L'enquete Criminelle*. 1931, Lyons, France: Joannés Desvigne et ses Fils.

4. Bock, J. H., The use of macroscopic plant remains in forensic science, in *The Encyclopedia of Quaternary Science*, S. A. Elias, editor. 2013, Elsevier: Amsterdam. pp. 542–7.

5. Simpson, B. B. and M. C. Ogorzaly, *Plants in Our World*. 1995, New York: McGraw Hill.

6. Sachs, J. S., *Corpse*. 2001, Cambridge, MA: Perseus Publishing. p. 270.

7. Bock, J. H. and D. O. Norris, *Forensic Plant Science*. 2016, Academic Press, London.

8. Bock, J. H., M. A. Lane, and D. O. Norris, *Identifying Plant Food Cells in Gastric Contents for Use in Forensic Investigations: A Laboratory Manual*. 1988, US Department of Justice, National Institute of Justice.

9. Norris, D. O. and J. H. Bock, Use of fecal material to associate a suspect with a crime scene: report of two cases. *Journal of Forensic Sciences*, 2000. **45**(1): pp. 184–7.

10. Moss not Grass, in *Forensic Files*, Season 10, Episode 29, https://youtu. be/K1YXdgF2eKM. 2006. 22 mins.

11. *Dominique Moss and Keith Lotmore v Regina*, in Criminal Appeal Nos. 11 & 14 of 2004, C.o.t.B.i.t.C.o. Appeal, Editor. 2004: https://www. courtofappeal.org.bs/judgments.php.

12. Miller Coyle, H. et al., Forensic botany: using plant evidence to aid in forensic death investigation. *Croatian Medical Journal*, 2005. **46**(4): pp. 606–12.

13. Aquila, I. et al., The role of forensic botany in reconstructing the dynamics of trauma from high falls. *Journal of Forensic Sciences*, 2019. **64**(3): pp. 920–4.

14. Aquila, I. et al., The role of forensic botany in solving a case: scientific evidence on the falsification of a crime scene. *Journal of Forensic Sciences*, 2018. **63**(3): pp. 961–4.

15. Hall, D. W., Case studies in forensic botany, in *Forensic Botany: A Practical Guide*, D. W. Hall and J. H. Byrd, editors. 2012, John Wiley & Sons, Ltd: Chichester, UK. pp. 174–87.

16. Cardoso, H. F. V. et al., Establishing a minimum postmortem interval of human remains in an advanced state of skeletonization using the growth rate of bryophytes and plant roots. *International Journal of Legal Medicine*, 2010. **124**(5): pp. 451–6.

17. Douglas, J. and M. Olshaker, *Journey Into Darkness*. 1997, New York: Scribner. p. 383.

18. Brabazon, H. et al., Plants to remotely detect human decomposition? *Trends in Plant Science*, 2020. **25**(10): pp. 947–9.

19. Carter, D.O., D. Yellowlees, and M. Tibbett, Cadaver decomposition in terrestrial ecosystems. *Naturwissenschaften*, 2007. **94**(1): pp. 12–24.

20. Hunter, J. R. and A. L. Martin, Locating buried remains, in *Studies in Crime: An Introduction to Forensic Anthropology*, J. R. Hunter, C. Roberts, and A. Martin, editors. 1996, B.T. Batsford Ltd: London. pp. 86–100.

21. Abdiu, S., Murdered man's body found after tree 'unusual for the area' grew from seed in his stomach, in *Daily Mirror*. 2018, MGN Limited: London.

22. Christou, J., Did a fig tree grow out of the remains of a Turkish Cypriot man missing since 1974?, in *CyprusMail Online* (available at: https://cyprus-mail.com/2018/09/23/did-a-fig-tree-grow-out-of-the-remains-of-a-turkish-cypriot-man-missing-since-1974/). 2018.

23. Caccianiga, M., S. Bottacin, and C. Cattaneo, Vegetation dynamics as a tool for detecting clandestine graves. *Journal of Forensic Sciences*, 2012. **57**(4): pp. 983–8.

24. Watson, C. J. and S. L. Forbes, An investigation of the vegetation associated with grave sites in Southern Ontario. *Canadian Society of Forensic Science Journal*, 2008. **41**(4): pp. 199–207.

25. Molina, C. M. et al., Geophysical and botanical monitoring of simulated graves in a tropical rainforest, Colombia, South America. *Journal of Applied Geophysics*, 2016. **135**: pp. 232–42.

26. Forensic Anthropology Center (https://fac.utk.edu/) (accessed 8 June 2020).

27. Dabbs, G. R., L. G. Roberts, and E. A. Zieman, There is a fungus among us, but is it useful for forensic analysis in human taphonomy?, in

Haglund and Sorg's Forensic Taphonomy: 21st Century Advances and Regional Variation, M. Sorg and W. Haglund, editors. Boca Raton, FL: CRC Press.

28. Jackson, S., *No Stone Unturned: The Story of NecroSearch International.* 2002, New York: Kensington Books. p. 374.

29. Bock, J. H. and D. O. Norris, Forensic botany: an under-utilized resource. *Journal of Forensic Sciences*, 1997. **42**(3): pp. 364–7.

Chapter 4

1. Langmead, C., *A Passion for Plants.* 1995, Oxford: Lion Publishing plc. p. 201.

2. Yates, N., *Beyond Evil: Inside the Twisted Mind of Ian Huntley.* 2005, London: John Blake Publishing Ltd. p. 292.

3. Wiltshire, P. E. J., Consideration of some taphonomic variables of relevance to forensic palynological investigation in the United Kingdom. *Forensic Science International*, 2006. **163**(3): pp. 173–82.

4. Khalil, K., The science of the Soham murders. *Medicine, Science and the Law*, 2005. **45**(3): pp. 187–93.

5. Wiltshire, P. E. J., Mycology in palaeoecology and forensic science. *Fungal Biology*, 2016. **120**(11): pp. 1272–90.

6. Wiltshire, P., Forensic ecology, botany, and palynology: some aspects of their role in criminal investigation, in *Criminal and Environmental Soil Forensics*, K. Ritz, D. Miller, and L. Dawson, editors. 2009, Springer Science + Business Media B.V, Dordrecht: Springer Netherlands. 2009. pp. 129–49.

7. Brown, A. G., The use of forensic botany and geology in war crimes investigations in NE Bosnia. *Forensic Science International*, 2006. **163**(3): pp. 204–10.

8. Mathewes, R. W., Forensic palynology in Canada: an overview with emphasis on archaeology and anthropology. *Forensic Science International*, 2006. **163**(3): pp. 198–203.

9. *Delgamuukw v British Columbia.* 3 S.C.R. 1010. 1997.

10. Siver, P., W. Lord, and D. McCarthy, Forensic limnology: the use of freshwater algal community ecology to link suspects to an aquatic crime scene in southern New England. *Journal of Forensic Sciences*, 1994. **39**(3): pp. 847–54.

11. Anonymous, *Crime Scene Creatures—Diatom Detective*. Public Broadcasting Service, USA. https://www.youtube.com/watch?v=mWQEB2_7-Tc (accessed 13 February 2020).

12. Scott, K. R. et al., Freshwater diatom persistence on clothing II: further analysis of species assemblage dynamics over investigative timescales. *Forensic Science International*, 2021. p. 326: doi: https://doi.org/10.1016/j.forsciint.2021.11089.

13. Webb, J. C. et al., Differential retention of pollen grains on clothing and the effectiveness of laboratory retrieval methods in forensic settings. *Forensic Science International*, 2018. **288**: pp. 36–45.

14. Pollanen, M. S., *Forensic Diatomology and Drowning*. 1998, Amsterdam: Elsevier. p. 159.

15. Pokines, J. T., and N. Higgs, Macroscopic taphonomic alterations to human bone recovered from marine environments. *Journal of Forensic Identification*, 2015. **65**(6): pp. 953–83.

16. Meachim, L., E. M. Grahame, and A. Mayes, Yacht carrying tonne of drugs hits Abrolhos Islands reef, alleged smugglers found on island, in ABC Midwest & Wheatbelt. 2019. ABC Australia. https://www.abc.net.au/news/2019-09-05/men-charged-as-alleged-drug-yacht-runs-aground-abrolhos-islands/11478908 (accessed 24 February 2022).

17. Pagliaro, E., *The green revolution: botanical contributions to forensics and drug enforcement. Part 2. Additional case studies*, in *Forensic Botany: Principles and Applications to Criminal Casework*, H.M. Coyle, Editor. 2005, Boca Raton, FL: CRC Press. pp. 179–84.

18. Trace Evidence: The Case Files of Dr. Henry Lee. Season 1, Episode 1. Hoeplinger & Matthison/Hawaii. 2004, KLS Communications Inc., Lawrence Schiller Productions. 43 mins.

19. Tedeschi, M., *Kidnapped*. 2015, Cammeray, NSW: Simon & Schuster (Australia) Pty Limited.

20. Locard, E., *Traité de Criminalistique: Vol 2 Les Empreintes et Les Traces dans L'enquete Criminelle*. 1931, Lyons, France: Joannés Desvigne et ses Fils.

21. Hawksworth, D. L. and P. Wiltshire, Forensic mycology: current perspectives. *Research and Reports in Forensic Medical Science*, 2015. **5**: pp. 75–83.

22. Wiltshire, P. E. J. et al., Two sources and two kinds of trace evidence: enhancing the links between clothing, footwear and crime scene. *Forensic Science International*, 2015. **254**: pp. 231–42.

23. *HMA v Patrick Rae*, J.o. Scotland, Editor. 2011: http://www.scotland-judiciary.org.uk.
24. Hawksworth, D. L. and P. Wiltshire, Forensic mycology: the use of fungi in criminal investigations. *Forensic Science International*, 2011. **206**: pp. 1–11.
25. Dabbs, G. R., L. G. Roberts, and E. A. Zieman, There is a fungus among us, but is it useful for forensic analysis in human taphonomy?, in *Haglund and Sorg's Forensic Taphonomy: 21st Century Advances and Regional Variation*, M. Sorg and W. Haglund, editors. Boca Raton, FL: CRC Press.
26. Hösükler, E. et al., Fungal growth on a corpse: a case report. *Romanian Journal of Legal Medicine*, 2018. **26**: pp. 158–61.
27. Babcock, D. W., The legal implications of 'toxic' mold exposure. *Journal of Environmental Health*, 2006. **68**(8): pp. 50–1.
28. Pettigrew, H. D. et al., Mold and human health: separating the wheat from the chaff. *Clinical Reviews in Allergy & Immunology*, 2010. **38**(2): pp. 148–55.

Chapter 5

1. Berry, W., Plant's DNA 'fingerprints' lead to man's murder conviction. 29, May 1993. Associated Press. Available at: https://apnews.com/326da16edd9e677ce6033700e470e9ae (accessed 19 January 2020). 1993.
2. Mestel, R., Murder trial features tree's genetic fingerprint. https://zephr.newscientist.com/article/mg13818750-600-murder-trial-features-trees-genetic-fingerprint/ (accessed 18 August 2021), in *New Scientist*. 1993, New Scientist Ltd.
3. Nybom, H., K. Weising, and B. Rotter, DNA fingerprinting in botany: past, present, future. *Investigative Genetics*, 2014. **5**(1): p. 1.
4. Arnaud, C. H., Thirty years of DNA forensics: how DNA has revolutionized criminal investigations. *Chemical & Engineering News*, 2017. **95**(37): pp. 16–20.
5. Combined DNA Index System (CODIS). Available from: https://www.fbi.gov/services/laboratory/biometric-analysis/codis 2020 (accessed 2 January 2020).
6. *State v Bogan*. 1995. 183 Ariz. 506 (1995) 905 P.2d 515.
7. Yoon, C. K., Botanical witness for the prosecution. *Science*, 1993. **260**: pp. 894–5.

8. *State v. Bible.* 175 Ariz. 549 (1993) 858 P.2d 1152.
9. Gustafson, D. J., D. J. Gibson, and D. L. Nickrent, Conservation genetics of two co-dominant grass species in an endangered grassland ecosystem. *Journal of Applied Ecology*, 2004. **41**(2): pp. 389–97.
10. Lee, C.-L. et al., Evaluation of plant seed DNA and botanical evidence for potential forensic applications. *Forensic Sciences Research*, 2020. **5**(1): pp. 55–63.
11. Craft, K. J., J. D. Owens, and M. V. Ashley, Application of plant DNA markers in forensic botany: genetic comparison of *Quercus* evidence leaves to crime scene trees using microsatellites. *Forensic Science International*, 2007. **165**(1): pp. 64–70.
12. Linacre, A., H. Hsieh, and J. C. Lee, Part III. Case study: DNA profiling to link drug seizures in the United Kingdom, in *Forensic Botany: Principles and Applications to Criminal Casework*, H. M. Coyle, editor. 2005, Boca Raton, FL: CRC Press. pp. 163–6.
13. Korpelainen, H. and V. Virtanen, DNA fingerprinting of mosses. *Journal of Forensic Sciences*, 2003. **48**: pp. 804–7.
14. Miller Coyle, H. et al., Tracking clonally propagated marijuana using amplified fragment length polymorphism (AFLP) analysis, in *Forensic Botany*, H. Miller Coyle, editor. 2005, Boca Raton, FL: CRC Press. pp. 185–96.
15. Shirley, N et al., Analysis of the NM101 marker for a population database of Cannabis seeds. *Journal of Forensic Sciences*, 2013. **58**(S1): doi: 10.1111/1556-4029.12005.
16. Phylos Galaxy project (https://phylos.bio/search/) (accessed 20 January 2020).
17. Ng, K. K. S. et al., Forensic timber identification: a case study of a CITES listed species, *Gonystylus bancanus* (Thymelaeaceae). *Forensic Science International: Genetics*, 2016. **23**: pp. 197–209.
18. Genbank: https://www.ncbi.nlm.nih.gov/genbank/ (accessed 20 January 2020).
19. Barcode of Life Data System: http://www.barcodinglife.org/. 2014–2021 (accessed 23 May 2021).
20. Pliny the Elder, *Natural History. Volume IV: Books XII–XVI.*, trans. L. H. Rackham. 1968. Cambridge, MA: Classical Literary, Harvard University Press.
21. Islam, N. et al., Toxic compounds in honey. *Journal of Applied Toxicology*, 2013. **34**: pp. 733–42.

22. Olive Genetic Diversity Database: http://www.bioinfo-cbs.org/ogdd/ (accessed 20 January 2020).

23. Ben Ayed, R. et al., OGDD (Olive Genetic Diversity Database): a microsatellite markers' genotypes database of worldwide olive trees for cultivar identification and virgin olive oil traceability. *Database*, 2016. **2016**: bav 090.

24. Mailer, R. J. and S. Gafner, Adulteration of olive (*Olea auropaea*) oil, in *Botanical Adulterants Prevention Bulletin*. 2020, Austin, TX: ABC-AHP-NCNPR Botanical Adulterants Prevention Program. p. 14.

25. O'Connor, A., Herbal supplements are often not what they seem, in *The New York Times*. 2013.

26. Newmaster, S. G. et al., DNA barcoding detects contamination and substitution in North American herbal products. *BMC Medicine*, 2013. **11**(1): p. 222.

27. Stoeckle, M. Y. et al., Commercial teas highlight plant DNA barcode identification successes and obstacles. *Scientific Reports*, 2011. **1**(1): p. 42.

28. *Monsanto Canada Inc. v Schmeiser*, [2004] 1 S.C.R. 902, 2004 SCC 34.

29. *Starlink v Aventis* 212 F. Supp. 2d 828 (N.D. Ill. 2002).

30. Schlessinger, L. and A. Endres, The missing link: farmers' class action against Syngenta may answer legal questions left after the StarLink and LibertyLink litigation. *Farmdoc Daily*, 2015. **5**: p. 35.

31. *Bayer Crop Science LP v Schafer, 10–1246 Ark 518 (2011)*.

32. Cohen, J., To feed its 1.4 billion, China bets big on genome editing of crops, in *Science Magazine*. 2019.

33. Ledford, H., CRISPR conundrum: strict European court ruling leaves food-testing labs without a plan. *Nature*, 2019. **572**: p. 15.

34. UNITE database https://unite.ut.ee/ (accessed 20 January 2020).

35. Nilsson, R. H. et al., The UNITE database for molecular identification of fungi: handling dark taxa and parallel taxonomic classifications. *Nucleic Acids Research*, 2018. **47**(D1): pp. D259–D264.

36. Young, J. M. and A. Linacre, Massively parallel sequencing is unlocking the potential of environmental trace evidence. *Forensic Science International: Genetics*, 2021. **50**: p. 102393.

37. Spencer, M. A., Forensic botany: time to embrace natural history collections, large scale environmental data and environmental DNA. *Emerging Topics in Life Sciences*, 2021. Sep. 24; 5(3): pp. 475–85. ETLS20200329.

38. Coghlan, S. A., A. B. A. Shafer, and J. R. Freeland, Development of an environmental DNA metabarcoding assay for aquatic vascular plant communities. *Environmental DNA*, 2021. **3**(2): pp. 372–87.
39. Watters, M. M., Fish and federalism: how the Asian Carp litigation highlights a deficiency in the federal common law displacement analysis. *Michigan Journal of Environmental & Administrative Law*, 2013. **2**(2): pp. 535–62.
40. *Michigan v U.S. Army Corps of Engineers* 667 F.3d (7th Cir. 2001). 2011.
41. Ishak, S., E. Dormontt, and J. M. Young, Microbiomes in forensic botany: a review. *Forensic Science, Medicine and Pathology*, 2021. **17**(2): pp. 297–307.
42. Jordan, D. and D. Mills, Past, present, and future of DNA typing for analyzing human and non-human forensic samples. *Frontiers in Ecology and Evolution*, 2021. **9**(Article 646130).

Chapter 6

1. Plato, *Plato in Twelve Volumes, Vol. 1* trans. Harold North Fowler; Introduction by W. R. M. Lamb; http://data.perseus.org/citations/urn:cts: greekLit:tlg0059.tlg004.perseus-eng1:117. Vol 1. 1966, Cambridge, MA: Harvard University Press.
2. Bruneton, J., *Toxic Plants Dangerous to Humans and Animals*. 1994, Paris: Lavoisier Publishing Inc.
3. Klaassen, C. D., ed. *Casarett and Doull's Toxicology: The Basic Science of Poisons*, ninth edition. 2019, New York: McGraw Hill Education. p. 1620.
4. Nepovimova, E. and K. Kuca, The history of poisoning: from ancient times until modern ERA. *Archives of Toxicology*, 2019. **93**(1): pp. 11–24.
5. Steenkamp, P. A. et al., Fatal Datura poisoning: identification of atropine and scopolamine by high performance liquid chromatography/photodiode array/mass spectrometry. *Forensic Science International*, 2004. **145**(1): pp. 31–9.
6. Hort, A., ed., *Theophrastus: Enquiry into Plants*. 1916, London and New York: William Heinemann and G.P. Putnam's Sons.
7. Aggrawal, A., Mass poisonings, in *Encyclopedia of Forensic and Legal Medicine*, 2nd edition, Vol. 3, J. J. Payne-James and R. W. Byard, editors. 2016, London: Elsevier Academic Press. pp. 300–6.
8. Lee, M. R., *Solanaceae IV: Atropa belladonna*, Deadly nightshade. *Journal of the Royal College of Physicians Edinburgh*, 2007. **37**: pp. 77–84.

9. Demirhan, A. et al., Anticholinergic toxic syndrome caused by *Atropa belladonna* fruit (deadly nightshade): a case report. *Turkish Journal of Anaesthesiology and Reanimation*, 2013. **41**(6): pp. 226–8.

10. Nathwani, A. C. et al., Polonium-210 poisoning: a first-hand account. *The Lancet*, 2016. **388**(10049): pp. 1075–80.

11. Nakagawa, T. and A. T. Tu, Murders with VX: Aum Shinrikyo in Japan and the assassination of Kim Jong-Nam in Malaysia. *Forensic Toxicology*, 2018. **36**(2): pp. 542–4.

12. Wennig, R., Back to the roots of modern analytical toxicology: Jean Servais Stas and the Bocarmé murder case. *Drug Testing and Analysis*, 2009. **1**(4): pp. 153–5.

13. Bock, J. H. and D. O. Norris, *Forensic Plant Science*. 2016, Academic Press, London.

14. Fornaciari, G. et al., A medieval case of digitalis poisoning: the sudden death of Cangrande della Scala, Lord of Verona (1291–1329). *Journal of Archaeological Science*, 2015. **54**: pp. 162–7.

15. Graeber, C., *The Good Nurse: A True Story of Medicine, Madness, and Murder*. 2013, New York: Twelve–Hachette Book Group. p. 213.

16. Health Care Professional Responsibility and Reporting Enhancement Act, in L.2005,c.83,s.1., T. S. o. N. Jersey, editor. 2005, Office of the Attorney General. p. 17.

17. Hardy, C. R. and J. R. Wallace, Algae in forensic investigations, in *Forensic Botany: A Practical Guide*, D. W. Hall and J. H. Byrd, editors. 2012, Chichester, West Sussex: John Wiley & Sons Ltd. pp. 145–73.

18. Desk, N., Alaska resident dies of paralytic shellfish poisoning (accessed 18 August 2020). Food Safety News 2020; available from: https://www.foodsafetynews.com/2020/07/alaska-resident-dies-of-paralytic-shellfish-poisoning/.

19. Tz'u, S. and B. E. McKnight, *The Washing Away of Wrongs: Forensic Medicine in Thirteenth Century China*. 1981, The University of Michigan, Center for Chinese Studies. p. 181.

20. Bretschneider, E., Botanical investigations into the *Materia Medica* of the ancient Chinese. *Journal of the North-China Branch of the Royal Asiatic Society*, 1895. **29**: pp. 1–623.

21. Zhou, Z. et al., *Gelsemium elegans* poisoning: a case with 8 months of follow-up and review of the literature. *Frontiers in Neurology*, 2017. **8**(Article 204).

22. Vickers, M., *Murder in California*. 2015, Larkspur, CA: Marquis Publishing.

23. Wolfe, S., Who killed Jane Stanford? (https://stanfordmag.org/contents /who-killed-jane-stanford) (accessed 16 May 2020), in *Stanford Magazine*. 2003.

24. Griffiths, G. D., Understanding ricin from a defensive viewpoint. *Toxins*, 2011. **3**: pp. 1373–92.

25. Schier, J. G. et al., Public health investigation after the discovery of ricin in a South Carolina postal facility. *American Journal of Public Health*, 2007. **97** **1**(suppl. 1): pp. S152–S157.

26. United States Department of Justice, New Boston, Texas woman sentenced for ricin letters, https://www.justice.gov/usao-edtx/pr/ new-boston-texas-woman-sentenced-ricin-letters (accessed 26 July 2021), O.o.P.A. Department of Justice, Eastern District of Texas. 2014 (updated 2015).

27. Pengelly, M., Envelope containing ricin was sent to White House, report says (accessed 21 September 2020), in *The Guardian*, London. 2020.

28. Shae, D. A. and F. Gottron, *Ricin: Technical Background and Potential Role in Terrorism*. 2013, Washington, DC: Congressional Research Service.

29. Audi, J. et al., Ricin poisoning: a comprehensive review. *JAMA*, 2005. **294**(18): pp. 2342–51.

30. Musshoff, F. and B. Madea, Ricin poisoning and forensic toxicology. *Drug Testing and Analysis*, 2009. **1**(4): pp. 184–91.

31. Papaloucas, M., C. Papaloucas, and A. Stergiolas, Ricin and the assassination of Georgi Markov. *Pakistan Journal of Biological Sciences*, 2008. **11**(19): pp. 2370–1.

32. BBC News. Alexander Perepilichnyy: the questions raised by Russian whistleblower inquest, https://www.bbc.com/news/uk-43767428 (accessed 27 February 2020). 2019.

33. Gomila Muñiz, I., J. Puiguriguer Ferrando, and L. Quesada Redondo, Primera confirmación en España del uso de la burundanga en una sumisión química atendida en urgencias [Drug facilitated crime using burundanga: first analytical confirmation in Spain]. *Medicina Clínica*, 2016. **147**(9): p. 421.

34. Ardila, A. and C. Moreno, Scopolamine intoxication as a model of transient global amnesia. *Brain and Cognition*, 1991. **15**(2): pp. 236–45.

35. Reichert, S. et al., Million dollar ride: crime committed during involuntary scopolamine intoxication. *Canadian Family Physician*, 2017. **63**: pp. 369–70.

36. West, N., *Encyclopedia of Political Assassinations*. 2017, Lanhan, Boulder, New York, London: Rowman & Littlefield Publishers.

37. Reijnen, G. et al., Post-mortem findings in 22 fatal *Taxus baccata* intoxications and a possible solution to its detection. *Journal of Forensic and Legal Medicine*, 2017. **52**: pp. 56–61.

38. Pilija, V., M. Djurendic-Brenesel, and S. Miletic, Fatal poisoning by ingestion of *Taxus baccata* leaves. *Forensic Science International*, 2018. **290**: pp. e1–e4.

39. Gaillard, Y., A. Krishnamoorthy, and F. Bevalot, *Cerbera odollam:* a 'suicide tree' and cause of death in the state of Kerala, India. *Journal of Ethnopharmacology*, 2004. **95**(2): pp. 123–6.

40. Menezes, R. G. et al., *Cerbera odollam* toxicity: a review. *Journal of Forensic and Legal Medicine*, 2018. **58**: pp. 113–16.

41. Stewart, M. J. et al., Findings in fatal cases of poisoning attributed to traditional remedies in South Africa. *Forensic Science International*, 1999. **101**(3): pp. 177–83.

42. Luyckx, A. V. et al., Adverse effects associated with the use of South African traditional folk remedies. *Central African Journal of Medicine*, 2004. **50**: pp. 46–51.

43. Lin, M.-Y. et al., Association of prescribed Chinese herbal medicine use with risk of end-stage renal disease in patients with chronic kidney disease. *Kidney International*, 2015. **88**(6): pp. 1365–73.

44. Akpan, E. E. and U. E. Ekrikpo, Acute renal failure induced by Chinese herbal medication in Nigeria. *Case Reports in Medicine*, 2015. **2015**: p. 150204.

45. Hawksworth, D. L. and P. Wiltshire, Forensic mycology: current perspectives. *Research and Reports in Forensic Medical Science*, 2015. **5**: pp. 75–83.

46. Wiltshire, P. E. J., D. L. Hawksworth, and K. J. Edwards, Light microscopy can reveal the consumption of a mixture of psychotropic plant and fungal material in suspicious death. *Journal of Forensic and Legal Medicine*, 2015. **34**: pp. 73–80.

47. Brown, A. C., Kidney toxicity related to herbs and dietary supplements: online table of case reports. Part 3 of 5 series. *Food and Chemical Toxicology*, 2017. **107**: pp. 502–19.

48. Krakauer, J., *Into the Wild*. 1996, New York: Doubleday. p. 207.

49. Krakauer, J. et al., Presence of L-canavanine in *Hedysarum alpinum seeds* and its potential role in the death of Chris McCandless. *Wilderness & Environmental Medicine*, 2015. **26**(1): pp. 36–42.

50. Krakauer, J., How Chris McCandless died. 2016 (cited 6 March 2020); available from: https://medium.com/galleys/how-chris-mccandless-died-992e6ce49410.

51. Aveline, J., The death of Claudius. *Historia: Zeitschrift fur Alte Geschichte*, 2004. **53**: pp. 453–75.

52. Wasson, R. G., The death of Claudius or mushrooms for murderers. *Botanical Museum Leaflets Harvard University*, 1972. **23**(3): pp. 101–28.

53. Javadzadeh, H. R. et al., *Citrullus colocynthis* as the cause of acute rectorrhagia. *Case Reports in Emergency Medicine*, 2013. **2013**: Article ID 652192.

54. Marmion, V. J. and T. E. J. Wiedemann, The death of Claudius. *Journal of the Royal Society of Medicine*, 2002. **95**(5): pp. 260–1.

55. Nordt, S. P., A. Manoguerra, and R. F. Clark, 5-year analysis of mushroom exposures in California. *The Western Journal of Medicine*, 2000. **173**(5): pp. 314–17.

56. Carus, W. S., *Bioterrorism and Biocrimes: The Illicit Use of Biological Agents Since 1900*. 2002, (Amsterdam, the Netherlands: Fredonia Books. p. 220.

Chapter 7

1. Hansen, E., *A Horticultural Tale of Love, Lust and Lunacy*. 2000, London: Methuen. p. 272.

2. Darwin, C., *Darwin Correspondence Project*, 'Letter no. 8773', accessed 8 June 2020, https://www.darwinproject.ac.uk/letter/DCP-LETT-8773.xml. 1873.

3. Bernstein, M., Man who smuggled endangered carnivorous plants into U.S. sentenced (https://www.oregonlive.com/portland/2017/08/man_who_smuggled_endangered_ca.html) (accessed 10 March 2020), in *The Oregonia/OregonLive*. 2017.

4. Brown, D., Planted (https://www.portlandmercury.com/feature/2017/07/05/19140283/planted) (accessed 10 March 2020), in *Portland Mercury*. 2017.

5. Goncalves, M. P. et al., *Justice for Forests: Improving Criminnal Justice Efforts to Combat Illegal Logging*. 2012, Washington, DC: The World Bank, p. 56.

6. Ohio Admin. Code 901:5-30-01 (https://www.registerofohio.state.oh. us/rules/search/details/294921) (accessed 24 February 2022). 2018.

7. Invasive Plant Species (https://www.registerofohio.state.oh.us/pdfs/ 901/5/30/901$5-30-01_PH_OF_N_RU_20171023_0826.pdf), S.o. Ohio, 2017.

8. Beaury, E. M., M. Patrick, and B. A. Bradley, Invaders for sale: the ongoing spread of invasive species by the plant trade industry. *Frontiers in Ecology and the Environment.* https://doi.org/10.1002/fee.2392, 2021.

9. Hoveka, L. N. et al., The noncoding trnH-psbA spacer, as an effective DNA barcode for aquatic freshwater plants, reveals prohibited invasive species in aquarium trade in South Africa. *South African Journal of Botany,* 2016. **102**: pp. 208–16.

10. Council, I. S. *Case Study: Mexican Feather Grass.* [pdf] November 2017 (accessed 31 March 2020); available from: https://invasives.org.au/wp-content/uploads/2017/11/Case-Study-Mexican-feather-grass.pdf.

11. Bunnings fined for selling noxious weed, in *Sydney Morning Herald* (https://www.smh.com.au/national/bunnings-fined-for-selling-noxious-weed-20100712-107sx.html) (accessed 31 March 2020). 2010.

12. Lamb, A. How to identify Mexican feather grass (https://www.youtube. com/watch?v=hUuw2a9HK3E) (accessed 31 March 2020). 11 February 2018; YouTube video.

13. Syme, A. E. et al., A test of sequence-matching algorithms for a DNA barcode database of invasive grasses. *DNA Barcodes,* 2013. **1**: pp. 19–26.

14. Convention on International Trade in Endangered Species of Wild Fauna and Flora (https://www.cites.org/eng) (accessed 29 May 2020).

15. United States Department of Justice. Lumber Liquidators Inc. sentenced for illegal importation of hardwood and related environmental crimes. 2016; available from: https://www.justice.gov/opa/ pr/lumber-liquidators-inc-sentenced-illegal-importation-hardwood-and-related-environmental (accessed 11 March 2020).

16. Lilongwe Wildlife Trust. New legislation offers lifeline to Malawi's forests. 2020; available from: https://www.lilongwewildlife.org/2020/ 03/02/new-law-offers-lifeline-to-malawis-forests/ (accessed 11 March 2020).

17. Environmental Investigation Agency, *The Rosewood Racket*. 2017, Washington, DC: Environmental Investigation Agency, Inc

18. UNODC, *World Wildlife Crime Report: Trafficking in Protected Species*. 2016, United Nations: Vienna. p. 97.

19. Dharmadasa, R. M. et al., Standardization of *Gyrinops Walla Gaertn.* (*Thymalaeaceae*): newly discovered, fragrant industrial potential, endemic plant from Sri Lanka. *World Journal of Agricultural Research*, 2013. **1**(6): pp. 101–3.

20. Nguyen, H. T. et al., Multi-platform metabolomics and a genetic approach support the authentication of agarwood produced by *Aquilaria crassna* and *Aquilaria malaccensis*. *Journal of Pharmaceutical and Biomedical Analysis*, 2017. **142**: pp. 136–44.

21. Lee, S. Y. et al., DNA barcoding of the endangered *Aquilaria* (*Thymelaeaceae*) and its application in species authentication of agarwood products traded in the market. *PLoS ONE*, 2016. **11**(4): p. e0154631.

22. Deng, X. et al., Characterization of the complete chloroplast genome of *Aquilaria sinensis*, an endangered agarwood-producing tree. *Mitochondrial DNA Part B*, 2020. **5**(1): pp. 422–3.

23. Slough, T., J. Kopas, and J. Urpelainen, Satellite-based deforestation alerts with training and incentives for patrolling facilitate community monitoring in the Peruvian Amazon. *Proceedings of the National Academy of Sciences*, 2021. **118**(29): p. e2015171118.

24. Taylor, R. et al., The rise of big data and supporting technologies in keeping watch on the world's forests. *International Forestry Review*, 2020. **22**(1): pp. 129–41.

25. Averyanov, L. V. et al., Field survey of *Paphiopedilum canhii*: from discovery to extinction. *Slipper Orchids*, 2014. **Fall 2014**: pp. 2–11.

26. Hinsley, A. et al., A review of the trade in orchids and its implications for conservation. *Botanical Journal of the Linnean Society*, 2017. **186**(4): pp. 435–55.

27. Gibson, D. J., Reviewed work: *Orchid Fever: A Horticultural Tale of Love, Lust, and Lunacy* by Eric Hansen. *Economic Botany*, 2001. **55**: pp. 173–4.

28. McClintock, J., Fakahatchee ghosts, in *Smithsonian Magazine* (https://www.smithsonianmag.com/science-nature/fakahatchee-ghosts-87585566/) (accessed 17 March 2020). 2003.

29. Orlean, S., *The Orchid Thief*. 1998, New York: Random House Publishing Group. p. 284.

30. *U.S. v Silva and Norris*, 04–20 144 CR-SEITZ (United States District Court Southern District of Florida). 2004.
31. Department of Justice, Peruvian orchid dealer sentenced to 21 months in Miami for smuggling protected Peruvian orchids, https://www.justice.gov/archive/opa/pr/2004/July/04_enrd_515.htm (accessed 14 August 2021), U.S.D.o. Justice. 2004.
32. *U.S. v Norris*, 452 F.3d 1275 (11th Cir. 2006).
33. Hinsley, A. et al., Estimating the extent and structure of trade in horticultural orchids via social media. *Conservation Biology*, 2016. **30**(5): pp. 1038–47.
34. Menteli, V. et al., Endemic plants of Crete in electronic trade and wildlife tourism: current patterns and implications for conservation. *Journal of Biological Research—Thessaloniki*, 2019. **26**(1): p. 10.
35. Wiedenhoeft, A. C. et al., Fraud and misrepresentation in retail forest products exceeds U.S. forensic wood science capacity. *PLoS ONE*, 2019. **14**(7): p. e0219917.
36. Lavorgna, A., Wildlife trafficking in the Internet age. *Crime Science*, 2014. **3**(1): p. 5.
37. McGivney, A., 'Yanked from the ground': cactus theft is ravaging the American desert (https://www.theguardian.com/environment/2019/feb/20/to-catch-a-cactus-thief-national-parks-fight-a-thorny-problem) (accessed 13 March 2020), in *The Guardian*.
38. Carver, E. and S. Ong, Tree sleuths. *Science News*, 2019. **196**(8): pp. 20–4.
39. Irwin, A., Cops and loggers. *Nature*, 2019. **568**: pp. 19–21.
40. UNODC, *Best Practice Guide for Forensic Timber Identification*. 2016, Vienna: United Nations. p. 226.
41. Kaldjian, E., L. Cheung, and M. Parker-Forney, 4 cutting-edge technologies to catch illegal loggers (https://www.wri.org/blog/2015/09/4-cutting-edge-technologies-catch-illegal-loggers) (accessed 15 March 2020), in *Insights*. 2015, Washington, DC: World Resources Institute.
42. Panshin, A. J., Comparative anatomy of the woods of the Meliaceae, sub-family Swietenioideae. *American Journal of Botany*, 1933. **20**: pp. 638–68.
43. Pastore, T. C. M. et al., Near infrared spectroscopy (NIRS) as a potential tool for monitoring trade of similar woods: discrimination of true mahogany, cedar, andiroba, and curupixá. *Holzforschung*, 2011. **65**(1): p. 73.

44. He, T. et al., Machine learning models with quantitative wood anatomy data can discriminate between *Swietenia macrophylla* and *Swietenia mahagoni*. *Forests*, 2020. **11**(1): p. 36.

45. Dormontt, E. E. et al., Forensic timber identification: it's time to integrate disciplines to combat illegal logging. *Biological Conservation*, 2015. **191**: pp. 790–8.

46. Agroisolab (https://www.agroisolab.com/) (accessed 31 May 2020). 2020. Agroisolab UK are experts in the use of isotope analysis for provenance testing, counter-fraud, and authentication of organic produce at the commercial level.

47. TIMTRACE Forensic timber tracing (http://www.timtrace.nl/) (accessed 31 May 2020). 2020.

48. DoubleHelix (https://www.doublehelixtracking.com/) (accessed 31 May 2020). 2020. Compliance and quality control for timber products.

49. Marc, J., How a library of tree DNA could protect the world's forests. 2019 [cited 15 March 2020]; available from: https://edition.cnn.com/2019/11/25/africa/gabon-tree-dna-library-scn-intl-c2e/index.html.

50. Jardine, D. I. et al., A set of 204 SNP and INDEL markers for bigleaf maple (*Acer macrophyllum Pursch*). *Conservation Genetics Resources*, 2015. 7(4): pp. 797–801.

51. United States Department of Justice, Mill owner sentenced to prison for purchases and sales of stolen figured maple from national forest (https://www.justice.gov/usao-wdwa/pr/mill-owner-sentenced-prison-purchases-and-sales-stolen-figured-maple-national-forest) (accessed 17 March 2020). 2016.

52. Double Helix. Plant DNA evidence supports landmark Lacey Act conviction of bigleaf maple theft. 2016. Available from: https://www.doublehelixtracking.com/blogs/2019/2/21/plant-dna-evidence-supports-landmark-lacey-act-conviction-of-bigleaf-maple-theft (accessed 17 March 2020).

53. Mundy, V. and G. Sant, *Traceability Systems in the CITES Context*. 2015, Cambridge: TRAFFIC International. p. 90.

54. United States Department of Justice, Essential oils company sentenced for Lacey Act and Endangered Species Act violations to pay $760,000 in fines, forfeiture, and community service, and to implement a comprehensive compliance plan (accessed 1 September 2020) (https://www.justice.gov/opa/pr/essential-oils-company-sentenced-lacey-act-and-endangered-species-act-violations-pay-760000). 2017.

55. American Herbal Products Association. Free primer on Convention on International Trade in Endangered Species (CITES). 4 June 2020. Available from: http://www.ahpa.org/News/LatestNews/TabId/96/ArtMID/1179/ArticleID/5/Free-primer-on-Convention-on-International-Trade-in-Endangered-Species.aspx.

56. Schmidt, B. M., Responsible use of medicinal plants for cosmetics. *Horticultural Science*, 2012. **47**(8): p. 985.

57. Department of Health and Human Services: Food and Drug Administration: notices; Matthew Dailey: final debarment order [fr doc 2020-05450], in 85. 2020. pp. 15193–4.

58. United States Department of Justice. Michigan man sentenced for unlawfully importing and distributing misbranded drugs. 2019. Available from: https://www.justice.gov/opa/pr/michigan-man-sentenced-unlawfully-importing-and-distributing-misbranded-drugs (accessed 23 March 2020).

59. Haber, I., J. Pergolizzi Jr., and J. A. LeQuang, Poppy seed tea: a short review and case study. *Pain and Therapy*, 2019. **8**(1): pp. 151–5.

60. Koop, C., Feds realize a jailed man's 3,000 pounds of 'marijuana' isn't pot, Texas lawyer says (https://www.msn.com/en-us/news/us/feds-realize-a-jailed-mans-3000-pounds-of-marijuana-isnt-pot-texas-lawyer-says/ar-BBYLr3w?ocid=spartandhp) (accessed 18 March 2020), in *Kansas City Star*. 2020.

61. Haney, A. and B. B. Kutscheid, Quantitative variation in the chemical constituents of marihuana from stands of naturalized *Cannabis sativa* L. in East-Central Illinois. *Economic Botany*, 1973. **27**(2): pp. 193–203.

62. Paris, C., Inside shipping's record cocaine bust (https://www.wsj.com/articles/inside-shippings-record-cocaine-bust-11563960604) (accessed 1 April 2020), in *The Wall Street Journal*. 2019.

63. Laboratory and Scientific Section, *Recommended Methods for the Identification and Analysis of Cocaine in Seized Materials (Revised and Updated)*. 2012, Vienna: United Nations Office on Drugs and Crime.

Coda

1. Conan Doyle, A., *The Adventures of Sherlock Holmes*. 1892, London: George Newnes Ltd. p. 307.

2. Wandersee, J. H. and E. E. Schussler, Preventing plant blindness. *American Biology Teacher*, 1999. **61**(2): pp. 82, 84–6.

3. Margulies, J. D. et al., Illegal wildlife trade and the persistence of 'plant blindness'. *Plants, People, Planet,* 2019. **1**(3): pp. 173–82.

4. Kane, S., *Seeds of Doubt.* 2004, New York: Scribner.

LIST OF CREDITS

Figures

Figure 1 Photo courtesy of Costa Rica Public Security Ministry

Figure 2 New Jersey State Museum https://www.nj.gov/state/archives/ slcspoo1.html. Photo courtesy of the NJSP Museum

Figure 3 Photo David Gibson

Figure 4 NJ State Archives https://www.nj.gov/state/archives/slcspoo1. html - fig 17. Photo courtesy of the NJSP Museum

Figure 5 Miller Coyle, H., et al., Forensic botany: Using plant evidence to aid in forensic death investigation. *Croatian Medical Journal*, 2005. 46(4): p. 606-612.

Figure 6 Figs 2 and 3 from Norris, D.O. and J.H. Bock, Use of fecal material to associate a suspect with a crime scene: Report of two cases. *Journal of Forensic Sciences*, 2000. 45(1): p. 184-187. Copyright 2000, ASTM International

Figure 7 Cardoso, H.F.V., et al., Establishing a minimum postmortem interval of human remains in an advanced state of skeletonization using the growth rate of bryophytes and plant roots. *International Journal of Legal Medicine*, 2010. 124(5): p. 451-456. Copyright 2009, Springer-Verlag

Figure 8 Reprinted by permission from Springer-Verlag: [*International Journal of Legal Medicine*] (Establishing a minimum postmortem interval of human remains in an advanced state of skeletonization using the growth rate of bryophytes and plant roots, H. F. V. Cardoso et al), Copyright © 2009

Figure 9 Brown, A.G., The use of forensic botany and geology in war crimes investigations in NE Bosnia. *Forensic Science International*, 2006. 163(3): p. 204-210. figure 1. Copyright 2006 Published by Elsevier Ireland Ltd

Figure 10 Hardy, C.R. and J.R. Wallace, Algae in Forensic Investigations, in *Forensic Botany: A Practical Guide*, D.W. Hall and J.H. Byrd,

Plates

PUBLISHER'S ACKNOWLEDGEMENTS

We are grateful for permission to include the following copyright material in this book.

Excerpt from Locard, E., *Manual de Technique Policiére*. 1923, Paris: Payot.

Excerpt from Mark Spencer, Forensic Botanist BBC. Plant detectives: how bramble and co. can help solve crimes. https://www.bbc.co.uk/programmes/articles/5q2xGXDZvoS7hg3KQl11vNg/plant-detectives-how-bramble-and-co-can-help-solve-crimes (accessed 27 January. 2020). 2020, courtesy of Dr Mark Spencer.

Excerpt from Langmead, C. (2000). *A Passion for Plants: The Life and Vision of Ghillean Prance*. Royal Botanic Gardens, Kew. © 2000 Board of Trustees of the Royal Botanic Gardens.

Excerpt from Mestel, R., Murder trial features tree's genetic fingerprint. https://zephr.newscientist.com/article/mg13818750-600-murder-trial-features-trees-genetic-fingerprint/ in *New Scientist*. 1993, New Scientist Ltd. 5 © 1993 New Scientist Ltd. All rights reserved. Distributed by Tribune Content Agency.

Excerpt(s) from *Orchid Fever: A Horticultural Tale of Love, Lust, and Lunacy* by Eric Hansen, copyright © 2000 by Eric Hansen. Used by permission of Pantheon Books, an imprint of the Knopf Doubleday Publishing Group, a division of Penguin Random House LLC. All rights reserved.

The publisher and author have made every effort to trace and contact all copyright holders before publication. If notified, the publisher will be pleased to rectify any errors or omissions at the earliest opportunity.

GENERAL INDEX

PLANT, ALGAE, AND FUNGAL SPECIES INDEX

Latin names (with their common names) of plants, algae, and fungi mentioned in this book are listed here. The names are those provided in the original sources plus updated synonyms. Plant common names are not standardized and can vary among sources and geographically, and in some sources only common names were provided in which case the Latin name is inferred. The abbreviation 'sp' indicates a single undetermined species in a genus, whereas 'spp.' indicates more than one species. In the text, Latin names are provided upon first mention of a plant in a chapter or part of a chapter with common names thereafter which are not indexed.